THEORY OF THE REARGUARD

THEORY OF THE REARGUARD

HOW TO SURVIVE CONTEMPORARY ART

(AND ALMOST EVERYTHING ELSE)

IVÁN DE LA NUEZ

Translated by **ELLEN JONES**

SEVEN STORIES PRESS

New York • Oakland • London

Seven Stories Press
140 Watts Street
New York, NY 10013
www.sevenstories.com

Library of Congress Cataloging-in-Publication Data

Names: Nuez, Iván de la, author. | Jones, Ellen, 1989- translator. | Nuez, Iván de la. Teoría de la retaugardia.
Title: Theory of the rearguard : how to survive contemporary art (and almost everything else) / Iván de la Nuez ; translated by Ellen Jones.
Other titles: Teoría de la retaugardia. English
Description: New York City : Sevens Stories Press, [2024]
Identifiers: LCCN 2024058729 (print) | LCCN 2024058730 (ebook) | ISBN 9781644214619 (trade paperback) | ISBN 9781644214626 (ebook)
Subjects: LCSH: Art and society--History--20th century. | Art and society--History--21st century.
Classification: LCC N72.S6 N8413 2024 (print) | LCC N72.S6 (ebook) | DDC 701/.03--dc23/eng/20250118
LC record available at https://lccn.loc.gov/2024058729
LC ebook record available at https://lccn.loc.gov/2024058730

College professors and high school and middle school teachers may order free examination copies of Seven Stories Press titles. Visit https://www.sevenstories.com/pg/resources-academics or email academic@sevenstories.com.

Printed in the United States of America

9 8 7 6 5 4 3 2 1

IN MEMORIAM

René de la Nuez

(San Antonio de los Baños, 1937–Havana, 2015)

CONTENTS

News 9

I. DUCHAMP AND THE DINOSAUR (How to Stop Art from Ruining a Good Doctrine)

1. When He Was Resurrected, the Readymade Was Still There 13
2. Don't Let Art Ruin a Good Doctrine 17
3. Platform? What Platform? 19
4. The Coming of the Non-Future 21

II. ART AS A POLITICS OF THE IMPOSSIBLE (How to Ride in My Limo Along Your Periphery)

5. La Lupe in the MoMA 25
6. A Franchise Called Contemporary Art 26
7. The Limousine and Its Metaphors 31

III. NEW VISUAL ORDER (How to Deal with Iconocracy)

8. Fear and Photography ... 41
9. From the Postmodern Condition to the Post-Photographic Condition ... 44
10. News of Photographic Antiquity ... 46
11. Iconocracy ... 51

IV. NEVER REAL, ALWAYS TRUE (How to Keep the Artist Doing the Moonwalk)

12. Artist Doing the Moonwalk ... 57
13. The Art Novel ... 60
14. Venice Without the Biennale ... 64
15. Museums, Muses, Musings ... 73

V. ONE OR THE OTHER—IMMORTAL OR CONTEMPORARY (How to Write an Epitaph for Contemporary Art)

16. Fukuyama, Contemporary Artist ... 81
17. The Art to Come ... 86
18. One or the Other: Immortal or Contemporary ... 89

Illustration Credits ... 96

NEWS

Fifty years ago, *Theory of the Avant-Garde* was published to immediate acclaim. In that book, its author, Peter Bürger, tackled what he saw as the two most important tasks of the artist: to break with representation and to dissolve the boundary separating it from life.

The failure of this dual mission proves, as Bürger himself admits, the collapse of the avant-garde and perhaps something worse: its impossibility.

Fifty years later, this book, *Theory of the Rearguard*, simply tests the waters of this great fiasco, though it wastes no time crying over it nor trying to paper over the cracks.

This is mainly because our times are marked not by the distance between art and life, but by a tension between art and *survival*, which is the continuation of life by other means. (More precarious means, to be sure.)

In the malaise born of that precarity, this book follows in the footsteps of an art that is leaving splinters in politics, iconography, and literature, fields from which it returns, increasingly worse for wear, to its perpetual Ithaca: the museum.

In all this back-and-forth, the "contemporary" is often used as a euphemism for seeking refuge in immortality. A refrain that has been repeating for a century so as not to have to face the end.

Under these circumstances, it's worth asking whether or not Contemporary Art will go on forever. If it is just as mortal as everything it invokes or examines under its magnifying glass, it would be worth writing an epitaph for it.

PART I

Duchamp and the Dinosaur (How to Stop Art from Ruining a Good Doctrine)

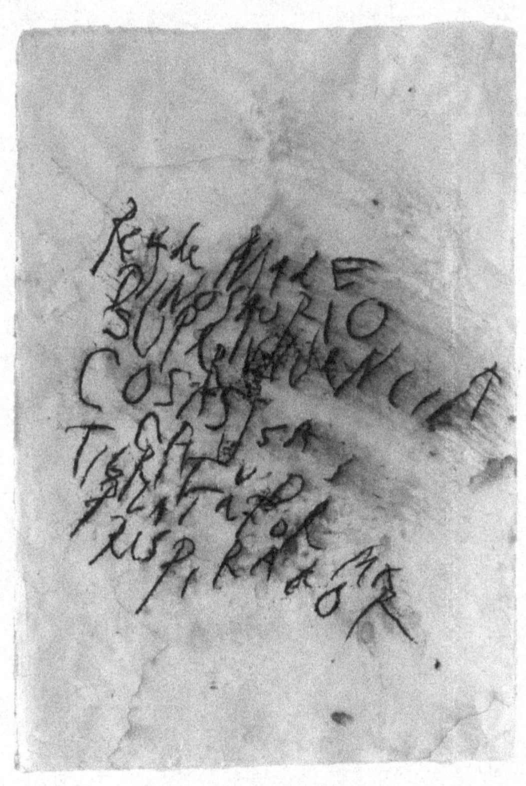

1. WHEN HE WAS RESURRECTED, THE READYMADE WAS STILL THERE

Every time Marcel Duchamp comes back to life, he finds that the readymade, like Augusto Monterroso's dinosaur, is still there. A readymade that, when Duchamp died in 1968, behaved like the avant-garde chapter of art or the blockhead revolution.

A fragment, in sum, of those higher classes governing history . . .

But now, every time the master completes the ritual of his return, he finds that, unlike when he was alive, it is those grand themes—art, the avant-garde, and revolution—that have been reduced to mere episodes in the history of an all-encompassing readymade.

This book positions itself against that colonization.

Against this adjectivized present in which anything can be recycled as "artistic." And in which, just as with that seminal urinal, things have moved, only now with the secret goal of neutralizing the subversive depth of their original meaning.

Whether they're in a gallery or in parliament, it makes no difference.

Welcome, then, to some of the landscapes in which the readymade becomes the definitive experience of this era in which everything, from the most sacred to the most profane, is already museum fodder: communism and the Spanish Civil War, the armed group Baader-Meinhof and Gaddafi's suits, Guantánamo and community welfare (as long as we consider it "one of the fine arts").

New technologies and old vanities come together to disseminate that continuation of the avant-garde by other means.

What does "by other means" mean? Well, that, just as war consists of a *violent* continuation of politics, as Clausewitz puts it, now art can operate as a *gentle* continuation of it.

And if Duchamp, or later Jeff Koons, made artistic entities out of particular *objects*—the urinal, a vacuum cleaner—by placing them in a gallery, now it is the *subjects*' turn.

First it was *things*, now it's *causes*.

As its penultimate turn of the screw, this ubiquitous readymade goes even further than exhibiting revolution or social battles, gender wars or injustices. We've reached the stage of exhibiting people.

And so, a museum in Malmö has exhibited two Romanian beggars.

Previously, in London, the playwright Brett Bailey took inspiration from the human zoos of the colonial era to show Black people in states of submission or domination. Further afield, the Jewish Museum of Berlin entertained us with "Jew in a Box."

In the same vein, there are those who have suggested turning the Guantánamo Bay detention camp into a museum . . .

It is not news that Contemporary Art resorts to exhibiting human beings, both dead and alive, with the objective of emphasizing exploitation, pain, or prostitution; with the intention of offering people a platform they do not have; or with the aim of shaking us out of our indifference . . .

The rejection of these methods used to come from Contemporary Art's enemies, but today there are many from that world itself who resist the idea of vacuum-packing social contradictions to serve them up later with supermarket logic—the fresh fish section, for example—that turns criticism into informing on people, speech into rhetoric, democracy into catharsis.

Needless to say, this is all done with the best intentions. Their canons are aimed at stereotypes and various forms of racism. And their goal is to shake up our Western consciences.

Nor need I insist that, if some of us find it hard to tell the difference between critique and frivolity, truth and image, culture and propaganda . . . it is our own lack of sensitivity that is to blame!

But the truth is that, by this stage, it's hard to endure these operations that denounce a crime by reproducing it, that ramp up domination to make it even more evident, and that go so far as to humiliate human beings . . . to establish the crisis of humanism!

All because of ignorance of the fact that "others," "subaltern subjects," or the "subdued" are as different from each other as those who pigeonhole them according to their presumptions.

Some time ago, fed up with these and other ruses, the Nigerian writer Wole Soyinka distanced himself from multiculturalism's blended negritude and opted instead for a *slightly* fiercer term: "tigritude."

The tiger does not go through the world worrying about defining himself. He simply acts according to his nature: He bides his time, leaps, devours you.

That's how you figure out he's a tiger. Too late, of course.

Perhaps we are witnessing the last throes of an aesthetic. Confirmation that the cycle begun by Duchamp has reached its endpoint.

In any case, this is no minor matter: If this decline is real, then it will no longer make sense to keep talking about what we lazily term Contemporary Art.

Just as it would make little sense to blindly follow Peter Bürger and his *Theory of the Avant-Garde*. Among other things, because the two challenges that art, in his opinion, was called upon to overcome—dissolving representation and therefore breaking the boundary that separates it from life—are revealed not only to be impossible tasks but also obsolete ones.

It would be more relevant, in the context of this defeat, to venture a modest theory of the rearguard. An exercise that would restore thinking about art, not in relation to life, but rather in tension with survival.

While I was editing the first draft of this book, I went back and updated my sources on the rearguard. I wanted to get away from its most common uses—constant military references—which either located "rearguardism" in a guerrilla context or treated it as a kind of mobile archive that, in different periods of history, held the memory of battlefronts.

This allowed me to establish how many more books on this topic are in English than in Spanish or any other language. And to establish, too, the term's drift towards the pornographic, with endless calls for anal sex (and its many invitations to "rearguard action" or "rear entry").

Portuguese sociologist Boaventura de Sousa Santos, in an interview published in 2018 by Spain's *El Salto*, and Mexican American critic Diana Taylor, in an interview published in 2016 by Ecuador's *El Telégrafo*, have confirmed the contemporary social and artistic scene's interest in the rearguard.

De Sousa proposes going beyond avant-garde ideas "capable of seeing what others do not see," which are supposedly ahead of their time. Instead, rearguard thinking would be capable of following, rather than generating, facts. A kind of thinking, in short, built for "those who are on the verge of giving up."

As for Taylor, in leaning towards the rearguard she crosses academic disciplinary boundaries to accept the importance of "the vulnerability of not knowing" when it comes to modelling an idea of the world.

I realized, with relief, that rearguardism, unlike vanguardism, is not a standardized term, and so the rearguard will never amass enough cultural capital to become an "-ism."

Whether I agreed or disagreed with these approaches, confirming the widespread interest in the topic made me feel like the rearguard is a place in which one is never alone.

The rearguard, that is, as a refuge for a resistance marked by the apogee of technology or the Age of the Image, by the idea that anyone can be an artist or the certainty that politics has become nothing more than performance, by the use of community work as a fine art or the abuse of "video-terrorism" as one of the dark arts, by art's place in fiction, and by the demagogy of models that continue to seek their causes outside of art while still benefitting from consequences only granted by the museum or the market.

2. DON'T LET ART RUIN A GOOD DOCTRINE

From that place in the rearguard, there's nothing better than avoiding the overlay of doctrine onto doubt, of the ideological map onto factual territory. And thereby sounding the alarm if you happen to drop into the Berlin Biennale, curated by Adam Szymczyk and Elena Filipovic, dedicated to exploring the ravages of gentrification and the advance of neoliberalism. An event ideally positioned to be critical of these issues and yet entirely incapable of examining modern Berlin without picking up on its remnants of fascism or communism, the SS or the Stasi.

And so, under the eclipse of this "curatorial strategy"—I solemnly swear not to use that term again for the remainder of this book—the event does an Olympic job of overlooking those two ideological giants, camouflaging itself as a sensible project that goes by the title *Cosas que no producen sombra* (Things that do not cast shadows).

Something similar happens when attending, in Madrid in 2013, an exhibition on 1980s Latin America in the Reina Sofía Museum. An exhaustive project, titled *Perder la forma humana* (Losing human form), that puts political art from the period center stage, and which is curated according to a set of theses put together by the Red Conceptualismos del Sur (Conceptualizations of the South Network).

At some point in the show, you try to find out the role of political art, artistic politics, anti-political art, and anti-artistic politics under the only two revolutions in power during that decade: the Cuban and the Nicaraguan revolutions.

And—don't let art ruin a good doctrine—you don't find anything. Those two revolutions have simply been amputated from the exhibition, which was very interesting until it had to confront that paradox.

In Berlin, they're reconstructing a modernity free of fascism or communism, free of the concentration camp or the Wall.

In Spain, they're reconstructing a Latin American radicalism that doesn't consider the revolutions that inspired it, nor the comrades to which they gave rise, for whom those revolutions were the air they breathed.

These absences do cast shadows.

Especially if we agree that Cuban art from that period was not very different from their neighbors' art in response to Operation Condor: collectively created, involving an anthropological dimension, political critique, the dismantling of national symbols, sexual liberation, the influence of Artaud, Grotowski, and Arte Povera, suspicion of expert voices . . .

And even more shadows are cast when we establish that in Nicaragua the opposite happened. That there, art driven by the Revolution portrayed itself as an archaic utopia, a return to the precolonial community along the lines of Ernesto Cardenal's social experiment in Solentiname, Nicaragua.

These revolutions are such a great counterpoint to "losing human form," as the Madrid exhibition was titled!

But paradox has little place in a Contemporary Art increasingly confused with the Gospel and which, the more it boasts of its appeal to knowledge, the more it fuses with faith.

In *On War*, Clausewitz recommends being slightly suspicious of causes, which justify everything, including our most lethal mistakes.

Pertinent critique should, instead, be established among the consequences. In that unglamorous area of war that takes refuge in the rearguard. Where the extreme forms of survival end up, the ones that do not find institutional counterparts to help successfully solve their new challenges.

3. PLATFORM? WHAT PLATFORM?

A theory of the rearguard knows that artists such as Duchamp or Beuys gave everything—or almost everything—to break that wall between art and life. But it also knows that this battle has not been exclusive to the avant-gardists. A Decadent such as Oscar Wilde advanced his own interests by bringing those two worlds together; and few have paid so high a price for that fusion.

And we haven't even got to G. K. Chesterton yet, who managed—through irony—to create a parable on art as anarchy in *The Man Who Was Thursday*. (This novel should be pressed into the hands of all the political art agencies currently multiplying around the world.)

A theory of the rearguard cannot fail to note the following coincidence, either: that the Kassel Documenta—so emphatic a reiteration of political art—and the French novelist Michel Houellebecq—who makes regular, fierce attacks on its collections—both called their antithetical projects *Platform*.

This was in 2008. And that similarity in name between a left-wing institution and a right-wing cynic leads us to consider things differently. On closer inspection, the platforms that best suit us might not be those that claim to be "programs," "agendas," or "strategies," but rather those that signal their physical meaning: as specific lifelines able to offer breathing space to survivors. To those who have moved between multiculturalism's zoological differentiation (a cage for every beast) and the absolute dissolution of the global standard (all beasts in the same cage, as long as they have lost their unique ferocity). Or to those who have shaken off real socialism and been overcome by unreal capitalism, trying to keep themselves afloat on their handful of supplies.

At this crossroads, knowledge or aesthetic emotion can only be conveyed at the cost of endangering the artist's own condition. And if the artist is already perceived—by Hegel, Agamben, or anyone else—as "lacking content" because their work expands "beyond" art itself then, we must rewrite our set of questions.

Is this a suicidal art we're seeing? If it were, wouldn't there be, as part of this dispossession, a desperate act of giving what you don't have? Is that not precisely what Lacan believed love to consist of—"giving what you don't have," as he puts it in *Trans-*

ference (translated by Bruce Fink)? And would this not point to the ecstasy of an art, with conceptualist pretensions, that turns out to be simply romantic?

Having delved into other worlds—politics, the media, technology, iconography, and literature—art leaves remnants of itself behind. That's why it behaves like a diminishing monster, returning each time worse for wear from its odyssey to The Institution and the range of prestige it confers.

This gap between an exultant departure and a humble return makes many of Contemporary Art's proposals not particularly credible, because the key lies not in its ability to overflow *beyond itself*, but in the fact that it cannot see through to the end the challenge posed by its expansion.

The reprehensible thing is not, as conservatives say, having strayed so far as to be out of one's depth, but rather the failure to convert that gamble into the measure of one's future.

4. THE COMING OF THE NON-FUTURE

What then, would be the future of art in a world that devotes itself every day to denying the future? Blanchot asked this question in his *Le Livre à venir* (*The Book to Come*, translated by Charlotte Mandell). And his answer was clear: It is precisely that lack of destiny that gives us the key to glimpsing tomorrow.

"Every art draws its origin from an exceptional fault."

So the art of a life with no future would have its advantages. One of them is that the artist—the writer, too, seeing as Blanchot makes no distinction between them—would arrive stripped of the desire to attain "the power and the glory." That detachment would even be enough to modify both the author's experience and that of whoever is receiving and questioning their creations.

Of course, Blanchot's future also predicted an atmosphere of "extraordinary turmoil that causes the writer to publish before writing, that causes the public to form and transmit what it does not understand, the critic to judge and define what he does not read, and the reader, finally, to have to read what is not yet written."

(Who could deny the importance of this master as an oracle!)

The point is that art, once situated in this present that was Blanchot's future, has at least three possible avatars awaiting it. One, to offer nostalgic irony for what was and can no longer be. Two, to consolidate its tendency to occur as *text* and to be enjoyed or despised as *reading material*. A third eventuality would operate on a more modest basis and would replace this period marked by a surplus of *possible* works with an era of *necessary* works.

On this point, rearguardist thinking would be defined by figuring out the relationship between art and survival. Precisely now, when we must contemplate not only the survival of art, but also the much more urgent art of survival.

It is at this precise moment that Marcel Duchamp comes back to life, washes his hands of the readymade, and utters his most precise self-definition: "I am a breather."

PART II

Art as a Politics of the Impossible (How to Ride in My Limo Along Your Periphery)

5. LA LUPE IN THE MOMA

"Politics is the art of the possible." Thus spoke Bismarck. As though positing the—Duchampian?—idea that we should accept politics as, at the very least, something "breathable."

The problem with that phrase is that it has a hard time staying afloat these days. Perhaps because two of the worlds that comprise it—art and politics—can themselves hardly boast of being *possible*.

During the successive borrowing between these two spheres, we can confirm that politics is becoming increasingly "aestheticized." Meanwhile, art is becoming "politicized," concerned with establishing administrative legitimacy, so to speak.

We can see that biennials as well as fairs, galleries, and museums are conscientiously applying themselves to the suturing of political issues, from the reunification of divided countries to dictatorships of various stripes, nationalism, cosmopolitanism, postcolonial strategies, and transitions to democracy . . .

And that's where we're all headed: curators, artists, critics, architects, and urban planners. Thanks to those great events and their noble discourses, we are assembling the aesthetic advance guard of a politics that needs to legitimize itself and an economy that needs to impose itself.

Faced with the dictates of the new economy, art and politics end up as ill-suited accomplices who deny their mutual contamination. As a result, it's not unusual for artists to acquire the worst defects of politics—rhetoric, cynicism, demagogy, messianism—which are added to those of art itself, particularly the much-discussed defect of "representation."

That "indignity of speaking for others," denounced by Foucault. The sustained performance that Agamben so disliked. The "pure theatre" that tormented La Lupe: "well-rehearsed falseness, studied simulation."

If politics is supposedly the art of the possible, Contemporary Art sets out to be a politics of the impossible.

A politics, though, that's closer to Bataille than to Bismarck, in which experience gleams as though teetering on the edge of an abyss, of a ravine in which the ghost of the possible lies in wait like a mocking spirit, vengeful of its death.

6. A FRANCHISE CALLED CONTEMPORARY ART

That specter flies mockingly over a system of art that is adjusting, insofar as it can, to the neoliberal climax that emerged from the ruins of the social-democratic model. That prototype it had formed with its institutions, public financing, language, and tacit agreement between "Hey you, violate me, but not too much" and "Hey you, buy me, but not too much."

Little by little, the new paradigm is growing with the steady pace of privatizations, shifts in prestige, and a vocabulary dedicated either to naming things or to embellishing them. We are obsessed with manufacturing metaphors, cultural industries, creative factories, the glorification of entrepreneurs. As though capitalism's chain of production—from the manufacture of products to their vacuum packing—could still boast a dimension that was subversive and unexplored.

Eventually, under the radical sheen of this new glossary, a semantics is reconciled, one that is closer than it might seem to the neoliberalism it is said to criticize. A neoliberalism that has mutated, by the way, and which today cannot be understood except as in political cahoots with the state. A kind of capitalism that has replaced competition with loyalty: in war as in peace, both here and in China, in Russia and Brazil, in Latin America's Bolivarian states and in the Arab Emirates in the Gulf.

From this perspective, the new era reaches an unusual magnitude in which a market Stalinism, imposed by the Right, is answered by a state Fordism, embraced by the Left.

The vertigo of this post-Fordism, under whose dominion we survive, is no longer satisfied with the self-exploitation today summed up as "do-it-(all)-yourself." Now, it also has to be dressed up with a narcissistic discourse best described as "talk-(only)-about-yourself."

All this has grown out of the fever for franchises that are making their way in contemporary culture, where they are to stay (and to reproduce). At this stage, not even the Louvre is exclusively a museum, nor the Sorbonne exclusively a university, nor Bertelsmann exclusively a publishing house. They are, first and foremost, global brands that—this is the thing about culture—have their own built-in gospel.

Think of a city like Barcelona and its decision to establish, right there in the port, a franchise of the Hermitage, the Russian art gallery that has been around since the time of the tsars, right through the Bolshevik Revolution, Stalin, two world wars, and the fall of communism all the way up to Putin. A museum that saw itself expanded, in tandem with the Guggenheim, to Las Vegas (a failure); and set up in Amsterdam, this time without joining forces (and with a degree of success).

After their frustrated bid for a Barcelona World mega-casino as an anti-crisis panacea, it's clear from their insistence on the Hermitage how attached authorities are to these franchises. From their insistence on a style that, beyond all the pretexts and euphemisms, puts branding before the city model.

This issue might be sparkling, but it isn't new. Twenty years ago, the US sociologist George Ritzer published *The McDonaldization of Society*, in which he warned of how the famous hamburger chain was colonizing other fields within the social organization of capitalism.

By means of parameters such as calculation, prediction, efficiency, and control—imported from the assembly line Ford began—Ritzer made it patently clear that this went beyond the delights of ground beef. No matter how varied the offerings—the touches of Mexican or Asian influence, the double cheese or veggie burgers—the McDonald's universe perpetuated itself as a strategic entity capable of standardizing consumer habits and behavior. (Needless to say, like any franchise, it was also religious about collecting its royalties.)

Franchises may not have an especially long history within capitalist culture, but they have certainly managed to establish a "tradition" that evokes the military base or the tourist enclave, colonialism or the crusades.

It's no surprise, then, that the Louvre, the Sorbonne, and the Guggenheim have disembarked in an emirate such as Abu Dhabi, where almost everything is yet to be done.

But what is surprising is that cities, or countries, with more complex cultural frameworks than the UAE should get involved in these kinds of ventures. And it's even more counterintuitive, if that's possible, at a time when Western art institutions—particularly European ones—are seriously depleted, both by the crisis and by the exhaustion of that social-democratic model we mentioned before.

Having guaranteed themselves, right there in the port, their dose of standard culture, will cruise ship passengers even set foot in the city's museums? Or will a port be different from an airport when it comes to handling a cultural offer whose essence rests, precisely, on neutrality?

In the absence of cultural models, our money goes to brands. In the absence of politics, we leave everything to finance. And in the absence of contemporary values, we cleave to those that seem perpetual (with their overdose of "classical" culture that guarantees queues at the museum doors when ticket sales are flagging).

And that's where we're at when the art world is suddenly enraged because three of its members aren't allowed into the UAE. The news confirms that they have been denied visas because they were critical of the working conditions of those constructing a museum complex like no other in Abu Dhabi.

One suspects these conditions differ little from those prevailing in the construction of other museums, universities, or hotels built or about to be built in these emerging landscapes.

But whatever . . .

The solidarity of the Gulf Labor Artist Coalition—and the solidarity shown towards it—is to be welcomed. Just as we would welcome it if the art world—its supreme and not-so-supreme leaders—were to look in the mirror, reflect, and, in the wake of such sad news, stop blaming all ills on distant landscapes or foreign entities. Just as we would welcome them recognizing something as simple as the fact that art does not live in the castle of purity, nor is it innocent of the process of expansion taking place in the global economy. Or if they were to accept that art is involved in the infamies of that model that it nonetheless goes on boldly condemning.

It's increasingly unsustainable to participate as the aesthetic avant-garde of this apogee of franchises while also clinging onto

Sartre and yelling about how "hell is other people." (As though golden retirements in the UAE were exclusive to retired footballers.)

The fact that three—or four or five or a hundred—of us aren't allowed in somewhere or other is a shame, but it's still a mere skirmish compared to the real battle that lies ahead. What's more, had they been let in—the three or four or five or a hundred of them—the impact of that irruption would still be minimal compared with the enormous interests agitating under the new aesthetic crusades. That negligible amount of power is, at the end of the day, all art is left with when it comes to these broader issues.

If the gentrification of cities such as New York, Berlin, or Barcelona occurred somewhat furtively, the gentrification taking place today in emerging territories is owing to the crude truth of oil money.

Without the illusion of independence or radical scenography to mitigate it, from now on even the most acrobatic of artistic curators will find it very difficult to keep their balance by putting their left foot in social revolt and their right foot in "petro-collections."

If it goes down this road, the art world will itself end up comprising a little emirate in which things that are banned in other contexts are permitted, with laws that differ even from those of other spaces it claims to represent.

It's enough to attend a biennial or an art fair to confirm that, just as "starchitecture" exists, so does "start," if you'll permit me the term, from the stars of the art world. Which goes to show that we are not the exception within the system but rather just another piece of its mechanism. An authorized display of its assembly line.

We're travelling in Don DeLillo's uncontaminated limousine, in David Foster Wallace's absurd cruise ship, a Hermitage or a Guggenheim waiting for us in the next port. We are, in short, another franchise called Contemporary Art, from which we validate capitalism's most savage practices while sublimating the theories of the naivest socialism.

If you have flicked through Blanchot, Jean-François Revel, and Toni Negri, or carried out serious projects around chapter 24 of *Capital*—yes, the one about original accumulation—you might think all this is a piece of cake and that what is written here does nothing but state the obvious.

It so happens, however, that this truth is not often assimilated. Out of naivety or cynicism, it's not clear which is worse.

Criticizing others also happens to be something of an experience, and that still offers some ideological consolation.

But, for the sake of its own credibility, that world needs to strengthen not the criticism of others but rather a profound self-criticism of art. An uncomfortable discipline that would have the value of holding a mirror up to its most opaque or blinding procedures.

7. THE LIMOUSINE AND ITS METAPHORS

Don DeLillo isn't the only one who rides in a limousine. For starters, there's David Cronenberg, who made a film adaptation of one of DeLillo's novels, the one about a corporate raider's journey through Manhattan with the sole aim of finally getting a haircut.

In his 1994 article, "The Coming Anarchy," Robert D. Kaplan used a limo to explain the emergence of global capitalism in the wake of the Cold War. This liberal thinker, whose diagnoses tend to be better than his cures, took a limo ride around the periphery and, from inside it, sent a burning question out into the world: Do all countries deserve democracy?

Kaplan's answer was categorical: Democracy is merely a "Western" accident, an ethnocentric superstition that, when it is taken to Asia, Africa, or even Latin America, becomes the kind of topic discussed in classrooms (but never in bars).

In the face of the future anarchy he foresaw, Kaplan showed a preference for the "benign" dictatorships of those years. Turkey, Peru, or China—countries able to guarantee economic growth and political stability with an iron fist—struck him as better options than the precarious democracies of countries then considered "ungovernable"—Russia, Nigeria, Colombia—where chaos had taken over. (Today we would call it disaster empowerment.)

To make his predictions, Kaplan appropriated a parable belonging to Homer-Dixon, a neo-Malthusian who also used the metaphor of the limousine to build an image of the world. (We're at four already!)

For Homer-Dixon and Kaplan, the difference between the limo and the neighborhoods through which it drove is much like the difference between the developed world and the peripheries. Or between countries that deserve a ride in democracy and those for whom no seat is reserved.

Needless to say, DeLillo's broker character, Cronenberg, Kaplan, and the ultraliberal Homer-Dixon all practice a version of financial Darwinism that considers it "natural" for some people to live in luxury while others are dispossessed.

And it's no coincidence that they have all used limousines, artifacts that can usually avoid the filth but rarely the traffic.

With the fall of communism, Contemporary Art, too, in its heyday, was tempted by a ride in one of those sets of wheels when it had to venture into more savage worlds.

Avoiding the filth while making the traffic worse.

Just think of its trips through Latin America.

When it comes to certain dilemmas that emerge from the

tearing down of the Berlin Wall, the circulation of ideas in Latin America has an advantage over Eastern Europe.

For starters, there they already know what the failure of the Left or the dismantling of the polarized world (*their* world, at least) looks like. They know what it means to transition to democracy and how the cocktail of neoliberalism and shock therapy is shaken.

They even know what it means to destroy, every day, a three-thousand-kilometer wall that marks the border—*the* border—with the United States.

These advantages mean they proceed with certain caution when it comes to gauging the impact of the Soviet Empire's collapse on this world across the Atlantic that Octavio Paz once called "the extreme West." (The West *in extremis*, more like.)

You only need to look at the two revolutions that had established power by 1989: Socialist Cuba and Sandinista Nicaragua. (The same revolutions that were cut from the exhibition titled *Perder la forma humana* [Losing human form] in order to prevent art from spoiling a good doctrine.)

OK, so the Nicaraguan Revolution—which was not governed according to the Soviet model—collapsed following elections and a civil war, three months after the Berlin Wall came down.

Cuba, though, which was structurally integrated into the imploding galaxy, managed to survive as a communist state alongside China, Vietnam, and North Korea.

This doesn't mean Latin America was a stranger to the Cold War: From the missile crisis in the Caribbean in the 1960s to the so-called low-intensity conflict in Central America in the eighties, by way of the dictatorships that took hold in the Southern Cone in the seventies, the battle for global hegemony between communism and capitalism also played out there. (Or rather, the battle between the Soviet Union and the United States did.)

But even taking this nexus into account, it would be wrong to say that communism's hecatomb had a mere knock-on effect in those regions.

Peculiarities visible in politics are even more pronounced in culture. November 9, 1989, found some Latin American intellectuals frantically trying to revive their canons, more worried about North–South relations than about the East–West conflict, and much more distressed by the crisis of modernity than by the crisis of communism.

What was being discussed in Latin America in 1989? Well... everything from the particularities of postmodernity to the repositioning of postcolonialism. From state repression to societal violence. From the validity of utopias to the effects of an anomalous modernity. From revision of identity determinism to inquiry into the status of tradition in contemporary society.

In the West Indies, these arguments were focused on going beyond the shadow of Caliban, who had monopolized depictions of island life.

To tackle these challenges, it wasn't enough to repeat the usual stance vis-à-vis the old colonial metropolises or US imperialism (important though it was to keep them in mind). That's where they diverge from the binary thinking of the 1960s, which, under the impact of the Cuban Revolution, had relied on liberation theology, dependency theory, and the aesthetics of the Boom.

The West was proclaiming its decadence once again? There was Nelly Richard to vindicate "the crisis of the original and the revenge of the copy." Or Roger Bartra to uncover the imaginary networks of political power that fed Mexican nationalism. Or Antonio Benítez-Rojo to emphasize a global Caribbean overspilling its own geography.

That's what we were discussing when Alberto Flores Galindo

recovered the Andean utopia on his journey—not in a limousine—to the origin of the Shining Path's violence in Peru, while Aníbal Quijano vindicated that utopia as a suitable way out, a way "of ceasing to be what we have never been."

Latin American culture had its own walls to pull down if it wanted to scale a different ladder; and it became clear that the great wall of stereotypes must also be destroyed if we want to rid ourselves of the colonizing gaze—both our own and that of others.

1989, that year when the wall comes down and Kaplan's limo starts to move around the world, is also the year of *Magiciens de la terre* (Magicians of the earth). This exhibition could be seen at the Pompidou Center a mere three months before the communist world began to be dismantled.

It would be absurd to establish a cause-and-effect relationship between the end of an exhibition and the end of an empire. But neither would it be wise to deny its impact on later projects.

Le Démon des anges ([The angels' demon] CRDC Nantes, Arts Santa Mònica Barcelona, Musée d'Art Contemporain de Lyon, 1989), *Kuba o.k.* (curated by Jürgen Harten, Kunsthalle Düsseldorf, 1990), and *Cocido y crudo* ([Cooked and raw] curated by Dan Cameron, Museo Nacional Centro de Arte Reina Sofía, Madrid, 1994) were three of the first exhibitions to strengthen the persistence of a peripheral (or subaltern) culture explained by First World institutions and curators, keen not to miss the multicultural limo and to load it with that mix of redemptive enthusiasm, criticism of the center by the center itself, exaltation of irrationalism, and the repertoire of exotic fantasies that Edward Said had already dealt with in *Orientalism* and *Culture and Imperialism*. All this without forgetting the standardization which, in the post-Berlin world, elided Latin America with Eastern Europe, Asia, and Africa, according to postcolonial codes.

1989 marks a link between the advent of postcommunism and the apogee of postcolonialism. Because, although communism had been defeated, it was certainly not dead and buried. Rather, it had been colonized by a capitalism that was kind enough to recycle its iconographic assets, its plutonium, its gas, its oil, and its authoritarianism.

When it comes to Contemporary Art, that diverse Latin America that welcomed the year 1989 was soon smoothed out by global standardization forcing it into overtly and covertly ethnocentric canons. Allocated as a cultural reserve for a West that has settled into the end of history. All at the cost of a reduction in Latin America's own diversity and a frequent denial of its ability to imagine itself.

Unlike colonialism or neocolonialism, postcolonialism has been presented as a correction, as a positive affirmation of *others*. Anyone can claim to be a specialist in postcolonial studies, but people think twice about calling themselves "postcolonialists." And anyone can show off their mastery of a subject, but nobody would talk about "postcolonizing" it.

Note that for years, criticism of this situation was not well received. Largely because of the deplorable argument that it was all done in the name of justice for subaltern cultures. And, as a result, because the large-scale triumph of multiculturalism managed to do away with all critical voices, hastily dismissing them as racist affronts.

But the crisis of multiculturalism is not only caused by reactionaries, learned or not-so-learned. It also obeys its own internal logic. And criticism of it comes not only from Donald Trump or the far-right wave that's building in Europe, but also from the cultures it claims to represent. In these cultures, there are people who refuse to operate as ethnic tropes, or to continually renegotiate their exoticism. Writers and artists who, in sum, have understood that decolonization is not a carnivalesque performance but rather a process that begins in your own mind, just as Frantz Fanon revealed in *The Wretched of the Earth.*

During the years in which multiculturalism was a great success in universities and biennials, anyone who dared contradict its overacting was silenced like an Uncle Tom who denied his own roots from inside his cabin. This was the era of the explosion of ethnic subjects (whose ethnicity was firmly attached), subaltern cultures (the cultures fixed firmly to the subsoil), cultural studies . . .

The irony is that these accusations often came from First World curators and academics, forever willing to provide insight while the Other is obliged to provide local color.

That these leaders had not yet been the object of laughter or ridicule can only be explained by the rampant multi-opportunism configured around their power, and the inadmissible truth that colonial manias infect every kind of ideology, including those on the Left.

The fact that Sting was unashamed at having paraded an Indigenous Amazonian "around the world" while swept up in the fervor of ethnic music, but was outraged when the man revealed he knew how to use a credit card and knock a few dollars out of him, says everything you need to know about this scourge. For Sting, an ethnic subject couldn't even be a picaro, because criminality is a Western defect outside the reach of the noble savage constructed by global music.

Today we know that, far from offering an alternative to globalization's standardization, multiculturalism has ended up functioning as its exotic phase. And if it undoubtedly found success in the rise of identity, it's also true that its failure lay in the free movement of diversity. It criticized, with good reason, the Uncle Tom in his cabin, but settled into any stand reserved for it at fairs, festivals, and biennials to ensure representation.

In the same way, it never placed peripheral cultures in the context of *their* modernity, because it was enough to encapsulate them in a time (the past of their ancestral condition) and a place (that of their national origin).

Perhaps it would have been fruitful for the followers of Said or Édouard Glissant to go back a little further in time, all the way to the promiscuity of Oswald de Andrade and Fernando Ortiz—with their metaphors of Brazilian anthropophagy and the rotten pot representing Cuba's chaos—instead of getting stranded in the exaltation of an infinite hyperanomaly that can only be defined by its critical inferiority to the West. That way they would have seen that multiculturalism was, in some cases, a step backwards with respect to "transculturation," a concept that Fernando Ortiz first put forward in 1940 and that would be worth updating today.

Another point to bear in mind—especially for the curatorial Left dictating much of art's destiny—is that multiculturalism reached its climax after the fall of communism. And that, in a way, its function has been to dilute, in culture, the ideological conflict faced by the polarized world. That was when the end of history went hand in hand with the end of geography, when the periphery was approaching the global centers.

If we must enter the jungle, Wifredo Lam's wouldn't be a bad start. And, once we're there, let's try and counter globalization's law of the jungle with the cultural opposition's law of the zoo.

In cages or enclosures, it's possible to keep things apart, but a mixture will never be achieved. Just like those limousines in which we traverse cities without ever getting to know them. Comfortably ensconced in a luxury wagon that allows us to photograph the world, but not understand its social imaginary.

PART III

New Visual Order (How to Deal with Iconocracy)

8. FEAR AND PHOTOGRAPHY

"When I am assailed by fear, I invent an image." Thus spoke Goethe. The only phrase necessary for us to understand *Ville panique* (*City of Panic*, translated by Julie Rose), a book by Paul Virilio about the relationship between terrorism and the urban experience. *Ville panique* complements *Ce qui arrive* (and the exhibition, *Unknown Quantity*), in which this theorist of the image tackles the topic of accident.

If, before Virilio, art imitated life, after Virilio, it imitated survival. And if before Virilio—under the aesthetics of the avant-garde—art could imitate politics, after Virilio—under the aesthetics of the rearguard—an assault can imitate nature.

Though their causes differ, accidents, natural disasters, and assaults have consequences in common: the sudden interruption of everyday life, the devastation of landscape, the ensuing trauma.

In these "panic" situations, the city abandons its old functions—the functions of modern life—and becomes a whale that detaches itself from old representations while swimming on towards other surfaces.

That's where it ceases to be a space of gathering and work and becomes one of atomization and idleness (or unemployment). It ceases to be an illusion of encounter or fulfillment and becomes instead a territory of loss and failure.

Seen thus, the utopias of More, Erasmus, Bacon, and Campanella can no longer explain the new dimension of this sperm whale describing the arc of its own failure.

We'll still call it "city," but really it's a "postcapital" entity, in the double sense of that term, because it both alludes to the sinking of its former function of representation (the city as the capital of a country, state, nation, community) and exists in a time when urban facts are marked by new economies in which the rhythm of capital, as with electronic music, is programed rather than produced; instead of reproducing, it simply begins to repeat itself.

What can be expected of images in a transition such as this? And what image would we have to invent, as Goethe did, in order to dispel our terror?

Might photography be the medium of fear?

Before answering, let's state the obvious: photography and the city haven't always gone hand in hand. Only in the modern era—the era of the apotheosis of both—does this correlation occur.

But although photography couldn't describe the founding of the city, it is an adequate medium for capturing its dissolution.

The same happens with definitions of photography itself. If there was a time when Walter Benjamin, Susan Sontag, Roland Barthes, and Virilio himself were speaking "about photography," today photography is able to speak for itself. And not only because of the emergence of the photographer turned theorist (Allan Sekula and Joan Fontcuberta) or the philosopher turned photographer (Jean Baudrillard). Rather, because photography is producing a *reflexive* image, an image with its own language, in the present wake of what Niépce called "sun writing."

From this perspective, photography manages to capture how the city is expelling what defined it during the twentieth century. And in this new condition, only rearguard art can find its place in this posterity of the city and of photography. This moment when the city—like a camera in the hands of a tourist—has ceased to be useful and has become simply "usable." (Like an old vinyl record in a DJ's suitcase.)

Obviously, many things in the city still work. But, precisely because of this, it's worth remembering that phrase so often used by photographers and soldiers and taken up by Paul Virilio in *La Machine de vision* (*The Vision Machine*, translated by Julie Rose): "If it works, it's obsolete!"

In this context, Joan Fontcuberta has announced the advent of a "New Visual Order." And he has done so based on a reflection on photography that is also a critique of it.

It's not that he repudiates photojournalists or documentary photographers. More than with photography, he takes issue with reality. And with the widespread perception that photography is a mere *reproduction* of it.

That's why the New Visual Order prefers to talk about truth than reality. Without being scared, as Godard would have it, of taking sides—"better a just image than just an image"—and with the conviction that every image should aspire to create an imaginary.

Ever since that May 1968 slogan—so overworked, so often betrayed—demanding power to the imagination, the New Visual Order reveals the power present in the image.

Something of this can be perceived in "Por un manifiesto posfotográfico" (For a post-photographic manifesto). In this article, Fontcuberta dissects the occupational hazards of a trade on the verge of disappearance. Though, curiously enough, it owes its disappearance not to its extinction but rather its *proliferation*.

The transformation of photography into a hobby, and of the camera into a human appendage (or even a nonhuman one, because there are now pets who take photos), has generated an unprecedented change in the photographer's place in the world, as well as in the images with which they try to narrate it.

"It is likely," Fontcuberta concludes, "that today Alonso Quijano would go mad not in libraries, devouring chivalric novels, but rather glued to the kaleidoscopic screen of a computer."

9. FROM THE POSTMODERN CONDITION TO THE POST-PHOTOGRAPHIC CONDITION

Between 1989 and 2001.

Between 11/9 and 9/11.

Between the Wall and the Twin Towers.

During this period, there is a shift from the aesthetics of disappearance (as Paul Virilio noted) to anxiety about the reappearance of the attack (as Virilio also outlined).

From one Virilio to another, the boundaries between collateral damage and "targeted" strikes are blurred. Between the weapons of mass destruction that were never found and the weapons of mass communication that will always be able to find us now, whether they belong to terrorists or those allied against them.

During those years, Fontcuberta was fine-tuning his ideas about the New Visual Order, a concept that reached its peak in 2015. The climax took place in Montreal, the city that had commissioned him to curate its photography biennial. Coincidentally, Montreal—specifically, the Council of Universities of Quebec—is the same city that, a quarter of a century earlier, had commissioned Jean-François Lyotard to write a "report on knowledge."

La Condition postmoderne (*The Postmodern Condition*, translated by Geoff Bennington and Brian Massumi) was the result of that brief, juicy study of postindustrial societies. An essay that activated all kinds of affiliations and adaptations to a great variety of contexts. From Chile to England, the United States to Mexico, France to India. But also from economy to art, literature to politics, the centers to the peripheries.

Let's just say that, twenty-five years later, Joan Fontcuberta closed the circle Lyotard had begun to trace, so his project could not have been titled anything other than *La condición postfotográfica* (The post-photographic condition).

A tribute to the French thinker and, at the same time, a visual reworking of his sociological theory. The end of an era and, at the same time, the beginning of another, one in which we are all photographers (or almost all of us, which is not the same, but similar enough, as the trova singer Silvio Rodríguez would say).

Stumbling through the works conceived in response to the idea of the "post-photographic condition," I began to wonder what a photographer *ought* to be doing these days. And I could think of no answer other than this: everything a tourist *cannot* do. So I'd better follow Marshall Berman's recommendation, read "the signs in the street," and assume that photography and artistic intervention involve the same action when it comes to dealing with the tension between congregation and dissolution, work and unemployment, home and eviction, tourists and immigrants.

An involvement, in short, as improvised as that of this city populated by ghost-like inhabitants; accomplices, victims, and witnesses to the violence exercised by the crowd on the community, by emptiness on solitude, by rejection on welcome, by the urbanite on the citizen.

10. NEWS OF PHOTOGRAPHIC ANTIQUITY

The New Visual Order had a past. And just as we have, according to Alexander Kluge, an ideological antiquity, we also carry with us a photographic "antiquity."

At the beginning of the twentieth century, photography was still an uncommon occurrence and the photographer's profession an unusual one. At the beginning of the twenty-first century, photography is a daily fact of life and the photographer's trade has been widely disseminated and thus, diluted.

In the Old Visual Order, photography was a promising novelty. In the New Visual Order, the photographic delirium has reached such dimensions that life is inconceivable without the millions of cameras adding, every second, to the overwhelming bank of images that this civilization will leave for those who come after us.

The Old Visual Order runs parallel to the twentieth century, which is the century of photography. A hundred years that encompass, for example, the life of a single photographer: Manuel Álvarez Bravo (who lived from 1902 to 2002).

This order belongs to the century of flash and to photographic light's conquest of world events.

The same light belonging to Robert Capa and Henri Cartier-Bresson (the "eye of the century").

Today we know that even certain photos considered historical reference points have been subject to manipulation and editing. Such is the case with Robert Capa's militiaman in the Spanish Civil War; the same goes for Dorothea Lange's concept of "photographic justice," which uncomplicatedly exalts propaganda.

But the twentieth century also gave rise to photographic honesty.

This was the case for Álvarez Bravo, who sees the photographer as a surreptitious actor, never a triggering one: someone who can, and must, remain safe.

Perhaps because photography should illuminate us not in the glare of the flash, but in the sense of providing knowledge. That's why his camera—*lucida*, as Barthes requested—often functions as a flashlight. For the photographer, for the photographic subject, and for the viewer.

Perhaps the problem with photography lies not in misunderstandings but rather in assumptions. And the fact is that, despite its appearance as the vanguard of modern technology, photography is a late art. We could say it has been delayed by at least seventy years. It should perhaps have been the art of the encyclopedia, the visual story of the Enlightenment. It should have illuminated the Enlightenment.

What's more, it should have been the art of revolution. A revolution painted by Jacques-Louis David is a postcard from classical antiquity, a "quotation" from Greece or Rome. Or an antecedent of Alexander Deineka and other painters of Soviet socialist realism, who dedicated themselves to showing us the perfect man we would one day be.

It's not that a painting cannot capture revolution, but that revolution doesn't look like much in a painting.

Wasn't this precisely why Mexican muralism took the walls by storm?

Let's compare "The Death of Marat" and the photo of Che lying dead, in the light of this disagreement between photography, revolution, and painting. Jacques-Louis David's painting is impregnated with all the photography that has not arrived yet. Freddy Alborta's portrait of Che carries within it all the painting we are already familiar with (it reminds us of "The Anatomy Lesson of Dr. Nicolaes Tulp"). Marat dies waiting for a photograph that is yet to be taken (though Charlotte Corday strikes him down quickly, with the sudden speed of a flash). Che, however, dies following a long wait that has a painterly result. (And that's despite having been the subject of the portrait by Alberto Korda, which has become the twentieth century's most reproduced photograph, also arranged, of course, as a Renaissance painting.)

In this sense, the Spanish Civil War and the Cuban Revolution had better luck than the French Revolution. (Better photographic luck.) The two former social processes appeared just when the photographic medium was ready to give them the visual coverage their respective movements required.

Agustí Centelles and Robert Capa, Walter Frentz and Heinrich Hoffmann, Osvaldo Salas and Alberto Korda, all combined war photography and propaganda, documentation and edulcoration, as both object and subject of those extreme processes of history—civil war, Nazism, revolution—that charged them with turning achievements into gestures.

Throughout history there have been statuary dictatorships and photographic dictatorships. The Stalinists, for instance, were eminently statuary tyrannies. Although ideology does not always define the method chosen by dictatorships to leave their mark on posterity. Mussolini was addicted to statues, while Franco can be defined as an equestrian tyrant.

Cuba is an example of the opposite. From his revolutionary beginnings, Fidel Castro could count on a large, well-qualified troop of what we might call battle photographers: Enrique Meneses, Osvaldo Salas, Raúl Corrales, Liborio Noval, Alberto Korda, Henri Cartier-Bresson, René Burri . . .

So he didn't need, like his "sibling" countries in the East, gigantic statues to spread the official image.

Photography turned out to be much a more modern and portable method of achieving this. And besides, it had an additional advantage: statues—of Stalin, Ceaușescu, Hussein—can be toppled; photographs cannot.

Henri Cartier-Bresson founds, Robert Capa builds, Alberto Korda embellishes. Manuel Álvarez Bravo, on the other hand, doesn't seem to have a plan. Cartier-Bresson and Capa want to *appear*, Álvarez Bravo prefers to *remain*. We might even say he prefers to *wait*.

And if he made photography the profession of an entire century—or, to be more exact, of only a single century—it's because he was a very unflashy photographer, as it were.

In a way, much of his work is to photography what pauses are to public speaking. It neither alarms nor is alarmed. As though always hiding something, as though always keeping something safe, just carefully *passing by*.

His photography is full of blushes, of striking gazes which, if we can grasp, we have already made our own. There is something primitive and caustic in it, a slight roughness.

The great thinkers of photography—from Susan Sontag to Paul Virilio, from Roland Barthes to Jean Baudrillard—often write from the perspective of the photographer or the camera. Manuel Álvarez Bravo demands another position: It's important, he says, to put yourself in the shoes of the photographic subject, no matter whether the subject is exceptional or, to use a frequent term, anonymous.

This is obviously why he looks down on photography as an act of violence. And it is also why it's so difficult to find another photographer who has picked his way so delicately through the minefield of others.

That territory indicates a temporal dimension more than a space. A time that Aurelio Asiain understands as a totality that stretches between the taking of the shot and what we eventually see (with the intervening shadow of the photographic developing process). A time that even extends into "something else" that the photographer cannot always identify but manages to show us; and that we eventually see, though we cannot always understand it.

Asiain also shows us a range of possible interpretations of Álvarez Bravo. All, in principle, acceptable; all, in the end, suspicious. One in which he's a surrealist and one where he's a social photographer, one where he's an artist and one where he's the precursor to magic realism; one where we glimpse the bearer of the nation's soul, and one that describes his picturesque flourishes . . .

This essay by Asiain contains many important ideas, but the most important is probably its catalogue of warnings about how *not* to read Manuel Álvarez Bravo. He reminds us that to stand before the photographer is to stand before "time passing." As well as before an arduous battle for the destruction of the pose.

We always get the impression that Manuel Álvarez Bravo gives out a warning before taking the shot, perhaps to give his photographic subjects, as in a children's game, enough time to hide.

Why? More out of necessity than humility.

Photography, for him, remains a social contract. A pact between individuals, an imaginary negotiation that takes into account both what is in front of and behind the camera. It's not about a subject and an object. These are two subjectivities that consciously interact.

It's as if he conceived photography as an act of normality in an anomalous world. And although he has photographed *characters*—some of them, frankly, quite important—his photography is stronger when he washes his hands of heroes. For all these reasons, he eschews the search for a protagonist that tends to characterize photography today. After all, we're not talking about a photographer of the exceptional—one bewitched by revolution or by war—but about a photographer of *people*, however warlike or revolutionary they may (or may not) be.

He is not interested in milestones, but in how the world evolves over time.

In many ways, the reason we write and make images today is down to the existence of photography in our tradition and experience—that visual antiquity of the Age of the Image.

That's why we don't even need to write *about* photography because our vocabulary is already *photographic*. We write essays and fiction, we create poems, out of a world in which language—including colloquial language—is marked by the camera.

Our words are forced, then, to compete with that light, and their survival involves digging a tunnel to escape it.

So maybe it won't be the ubiquitous glare of photography that will generate the future's best texts, but rather outright photophobia.

11. ICONOCRACY

What Joan Fontcuberta defines as the New Visual Order, I like to refer to as "iconocracy." A term that, in principle, points to the image dictatorship but which also functions as an ecosystem of power and counterpower.

Iconocracy appeals, of course, to the images of that power, but also to the power of images and, we must never forget, to their vengeance.

In this sense, more than iconoclasm, the new imaginary seems to entrust itself to "iconophagy," a term shared by Norval Baitello Junior and Alfonso Morales, and first introduced by Oswald de Andrade and Fernando Ortiz Fernández. A process of gestures and digestion that appeals to somewhere other than the photographer's gaze and to a radical removal of what is usually thought of as a photographic object.

While in the time of the Old Visual Order, avant-garde artists came to situate art within the prescriptions of agitprop, in the New Visual Order, artists—from the rearguard—have managed to dissolve agitprop in the rules of art.

We're talking about producers of visual imaginaries who have settled into a period designated the Age of the Image, in which visual culture has begun to replace written culture as a source of knowledge transmission. What happens is that images take on the role of creating new rhetorics and bringing center stage figures other than the one known, in previous times, as "the intellectual."

Seen thus, iconocracy deals with images as *knowledge.* But it also deals with them as *power.*

Are our lives not ruled by icons? We wake up, and an icon turns the world on. We go to sleep, and another turns it off. Whether we want to figure out the time or where we are in space, it makes no difference: Everything involves clicking on the correct symbol. If someone makes something of themselves in this world, they no longer become a hero or a martyr, but rather an icon, the ultimate confirmation of contemporary animism.

With the collapse of communism—and with the passage from one PC (Partido Comunista [the Communist Party]) to another (the Personal Computer)—the Digital Age–cum–Global Age–cum–Age of the Image–cum–New Visual Order was decreed. Any of these denominations confirms the assault on power by images, with that proliferation of selfies, photos, and visual networks from which it is impossible to escape. (This is the moment when the "camera lucida"—as invoked by Barthes—becomes the camera *narcissista.*)

Now, if this is the Age of the Image, what, then, would be the image of this age? Which of all the images in existence would qualify as the icon that defines it? Which one, at the end of the day, has the inarguable stamp that proves it is worth more than a thousand other images?

There are images of the ruins of the Berlin Wall and of the Twin Towers. Photos of street protests or the facial composite of *The Protester* polished up for the cover of *Time* magazine. There are images of the wars lingering in the aftermath of the Cold War and the occasional picture of a city after an attack. The millions of daily selfies and indescribable photos of tourists taking photos.

And there is the photo of a boy, Alan Kurdi, dead on the beach. That tragedy cut from an enormous picture that includes millions of displaced people. (And which alone is enough to personify this culture's malaise.)

All these are photojournalism images, the kind you see at events such as Visa pour l'Image or World Press Photo. Images in which photography dovetails with reality.

But we must also consider iconographies that do not "illustrate" or "raise awareness" (of a catastrophe, a record, a conquest), but that actually show how hard it is to understand what is going on. This critique of images for images' sake has been called iconophagy. The contemporary history of which can perhaps be linked to Artaud's demand that we never be real and always be true.

Faced with an event as thoroughly visualized as the attack on the Twin Towers, Thomas Ruff set about applying this ultimatum. It's important to describe his strategy for complicating the perception of the 9/11 massacre in New York. From a normal distance, we can see the building burning; everything is clear. However, as we get closer, the photograph becomes more pixelated, out of focus, leaving us with a blurry image: a perplexing abstract landscape of everything that happened.

It turns out, in the end, that the closer we get, the less we see.

While Stockhausen came to define the attack as the perfect work of art, Ruff is not interested in the perfection of the horror, but in the intrinsic obstacle to understanding it. Where Stockhausen *sees*, Ruff perceives the ignorance of those who did not get to see. Leaving a question hanging over us about a destruction that we have not fully understood, but which still requires us to *believe*.

In that faith lies the same manipulation of the images—whether just one or a whole cascade of them—that mark this age.

If to speak is to lie, as Foucault warned when he spoke of the traps of language, the same could be said about the image.

To photograph is to lie.

Or, sometimes it is. At least, it also is.

PART IV

Never Real, Always True (How to Keep the Artist Doing the Moonwalk)*

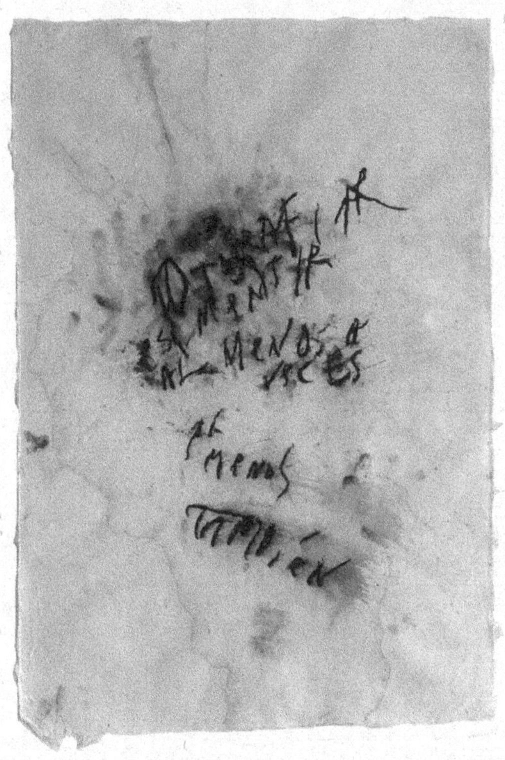

* "Never Real, Always True" served as the conceit for the group exhibition *Nunca real / Siempre verdadero* at Azkuna Zentroa–Alhóndiga Bilbao from March 14, 2019 to September 22, 2019.

12. ARTIST DOING THE MOONWALK

Since we're already investigating Old and New Visual Orders, it's worth scrutinizing a moment in Contemporary Art that nobody has stopped to consider. That moment reaches its peak in the 1980s. We could say it coincided with the crisis of communism, the explosion of the internet, and the promotion of the Global Era.

Around the time of these events, a strange way of narrating artists' biographies was agreed upon. Suddenly, the résumés in catalogues began to be written backwards. That is, from the present to the creators' origins.

2018

2017

2016

2015 . . .

And so on until, let's say, 1978.

With this reverse movement, the story art tells about itself suffers a temporal upheaval that reveals its horror of the future.

Thanks to this retreat, the artist's résumé is turned into a Freudian artifact, a return journey to the mother's womb.

It begins in the present, but quickly retraces its steps to the protagonist's real or artistic date of birth.

That "reverse" résumé not only connects to Freud. It also connects to fiction (much more so than to criticism or theory). It's closer to F. Scott Fitzgerald's "The Curious Case of Benjamin Button" or Alejo Carpentier's "Viaje a la semilla" ("Journey Back to the Source," translated by Frances Partridge) than anything by Arthur Danto or Brian Holmes.

Both tales depict an escape into the past from a present obsessed with turning its back on the future. Thanks to this narrative tempo, the résumé is no longer a reliable tool because its mission is none other than to betray biography.

While the résumé privileges awards, the biography feeds on more lurid material. If it wants to fulfil its objectives, a résumé must *conceal*; if it is honest, a biography must *reveal*. In the face of the résumé's professional asepsis, all the vices, obsessions, vanities, and grudges that populate the biography, that novel of art—which has got longer and longer in recent years—rise to the surface. As is evident from any catalogue, the résumé retreats while life moves forwards. Sailing hopelessly towards a horizon where uncertainty, decrepitude, and our end await us.

Since the birth of the image—later assigned to a figure named the "artist"—human beings have tried to leave proof that they have "produced something" in their lifetime—or as we say in Spanish, "pintado algo" (painted something).

It is worth pausing for a second on this commonplace, which equates "existing" with "painting." Or "art" with "life," as the old avant-garde used to say, which may too have failed because of the sublimation of one of the conflicting parts: art. And because of the observable fact that for avant-gardists, life—minuscule and finite—has always mattered little. "Siempre ha pintado poco," as the saying goes—it's never made much of a mark.

Seen thus, both the staunch defenders of art as an eternal entity and its professional gravediggers are wasting their time. Because, while it's true that no decree can make art disappear, it's also true that no decree can guarantee its immortality. Should art continue, it seems unlikely it will manage to do so by running backwards in time, with the artist seeking shelter in neonatal purity. And if it turns out to be nothing more than a transitory trade, like so many others in this world, we will only find out by setting the clock to run forwards: by marking its time in parallel with our own mortal calendar, heading into the future.

According to the chronology of résumés, art can only give us, as José Lezama Lima pointed out, a partial epiphany.

With that backward movement—that devious moonwalk that simulates an advance—art refuses to invoke progress in favor of simply *representing* it. And if it persists as a remote trade that refuses to capitulate, it's not because its story is realer or truer than that of other occupations that have disappeared over the course of history. Rather, it's mainly because it can still boast of being *exceptional*. Because it is still able to simulate lifestyles (political, hedonist, economic) that are forbidden in other worlds.

Art's problem, then, doesn't lie exclusively in the fact it cannot conjure up the horizon. Just as serious, if not more so, is the fundamental displacement of its vocabulary. That lingua franca separating itself from art's mother tongue to lodge itself instead in the new dictionary of art's display (which is slightly different).

A conservative constant rings out amid all this. At the end of the day, as Oscar Wilde anticipated, we do not return to our origins because the past was better, but because we cannot change it.

It's not the quality of the past, but its certainty that leads us to extoll it.

Can there be any greater guarantee of preservation than a return to the known? And can there be any greater guarantee of survival than the choice of a known *reality* over an as yet unknown *truth*?

13. THE ART NOVEL

"Never real and always true." Thus spoke Artaud. And thus he wrote, on a drawing—this compressed manifesto on the value of art as truth and, at the same time, on the scant virtue of relating it to reality (or what we assume to be reality).

Never, always . . .

Resounding magnitudes broken by more ambiguous scales, such as art and writing, truth and reality, future and past . . .

All this is set to collide, in 2013, in Kassel, a city where Documenta has taken place every five years since 1980.

And in Kassel there is an appropriate vantage point from which to watch this collision of worlds: the Artaud Café, no less.

From that very vantage point, Enrique Vila-Matas interrogated the mysteries of Contemporary Art in his novel *Kassel no invita a la lógica* (*The Illogic of Kassel*, translated by Anne McLean and Anna Milsom), in which he details his mishaps as a "guest artist" at Documenta 13.

It isn't the first time Vila-Matas has targeted the art world in his writing. In a 1991 story—"Me dicen que diga quién soy" ("They Say I Should Say Who I Am," translated by Margaret Jull Costa)—he gave an account of Panizo del Valle, a multicultural artist who shamelessly fused commitment and colonialism, success and cynicism. (The character wouldn't be out of place in a Renzo Martens documentary.) Between the story and his more recent novel, we might even be able to establish a history of Contemporary Art as a theme in fiction.

This isn't to say that the Barcelona novelist was the first—nor the last—to decide to turn art into a literary genre. In the distance we still have the foundational works of Oscar Wilde and Henry James, G. K. Chesterton and Gabriele D'Annunzio. A little closer, we have Guy Davenport, Aldous Huxley, and Marc Saporta. And these days, there's Paul Auster and Don DeLillo, Patrick McGrath and Michael Cunningham, Michel Houellebecq and Siri Hustvedt . . .

Donna Tartt won the Pulitzer in 2014 for *The Goldfinch*, a novel that begins in the Metropolitan Museum of Art in New York, the same place where, thirty years earlier, in *Wittgenstein's Mistress* by David Markson, the last living being on Earth seeks shelter.

In Ibero-America, where art has been, for fiction, a more extemporaneous than contemporary phenomenon, we have Julián Ríos and Ignacio Vidal-Folch, Álvaro Enrigue and César Aira, Julián Rodríguez and Agustín Fernández Mallo, Juan Francisco Ferré and Alicia Kopf.

The variety of writers here shows that art's presence in contemporary novels does not respond so much to a style, current, approach, or quality, but to the irrevocable visual flood that has managed to inundate everything, including literature.

In any case, name dropping does more harm than injustice, so I'd better stop the torrent of names which, over the last two decades, have managed to fatten that imaginary collection destined not for the museum but for the library. And which proposes an art made not for spectators but rather for *readers*.

This "authors'" collection not only focuses on artists (imagined or otherwise) and their work (imagined or otherwise). It also involves museums (Orhan Pamuk), directors (Ignacio Vidal-Folch), collectors (Steve Martin), and social art projects (Miguel Ángel Hernández Navarro). It even has a muse, Sophie Calle, who has captured the gaze of Paul Auster, Grégoire Bouillier, and Enrique Vila-Matas himself.

The reasons for this fascination, then, are as diverse as their authors. And in general, it is also positive: both for the enrichment of literary plotting and for the examination of certain less obvious mechanisms of art.

If you want to know what an artist *does*, it's enough to visit galleries and museums. But if you want to know who an artist *is*, you should read these novels.

Because, while museums will continue to praise their string of successes, these fictional pieces bring us closer to their failures, their miseries, the slips in their trajectories.

Biographical richness is not the only key to understanding the growing success of this "artistic" literature. (An increase in presence is not enough to prove the existence of a subgenre, nor does it fully explain the current contamination between these two spheres that once kept a hostile distance.)

Recent encounters speak to us of an art and a literature that are both perceived to be in extremis, typical of this Age of the Image in which writing no longer has a monopoly over the array of our knowledge, but art *still* has not managed to take over the task completely.

So this promiscuity ends up being "natural," necessary, and inevitable rather than extravagant.

While the artistic incursions of Oscar Wilde or Henry James were exceptional in the nineteenth century, those of twenty-first century writers are beginning to function as a rule. (Sometimes—the horror!—as a fashion.)

And if nineteenth-century novels about art privileged creative torment, bohemian lifestyles, and the romantic halo around the artist's condition, in the twenty-first century the range extends to the possibilities offered by the document and the archive, the artistic process and critical traps, collecting and the museum, directors and curators, politics and money, bodily changes and the emergence of technology.

And while in other times it was enough to write *about* a phenomenon, now the best novels are those that speak to us *from* a phenomenon. When art appears in César Aira's prose, it isn't just as another plot element to be described (in the same way as a landscape, action, or character is) but rather as a narrative resource that sustains the architecture of his books. And when Jeff Koons and Damien Hirst make an appearance in Michel Houellebecq's *La Carte et le territoire* (*The Map and the Territory*, translated by Gavin Bowd), it's not so much to emphasize how important they are but rather because the use of their real names allows the contemporary usefulness of something as overworked as a roman à clef to be brought into question.

In both cases, the presence of art in the novel has the effect of *broadening the battlefield.*

Through this broadening, art fiction allows us to open ourselves up to other possibilities that have traditionally been denied art criticism. So it is that Arthur Danto's theories would struggle to surpass Steve Martin's when it comes to scrutinizing the last twenty years of New York's art scene. (From the Chelsea boom to the bursting of the real estate bubble by way of the Japanese invasion, the emergence of Chinese art, and the collapse of the Twin Towers, as we read in Martin's *An Object of Beauty*.)

On the other hand, it's impossible to avoid the fact that the massive irruption of art in literature has climaxed right when art has lost its aura. So what most of these books take for granted is not the power of art but rather its vulnerability. It's not art's strength, but rather its fragility that has increased its presence in fiction. (And there is nothing more captivatingly novelesque than a fragile hero or a fallen angel.)

If one of art's recent goals has been to move beyond itself—we're back to Hegel—with a view to extending its reach to politics and social action, one of its chronic problems has been its inevitable return to the museum. To those safe quarters that have continued to function as the definitive sphere of its gratification.

Art fiction, on the contrary, has the advantage of being able to keep art in that space into which it has expanded—like Roberto Calasso's gods trapped mercilessly in literature—and prevent it from returning, safe and sound, to its lifelong Ithaca.

14. VENICE WITHOUT THE BIENNALE*

Realizing this, the cleverest artists plot their revenge by establishing their own way of narrating, with codes that even the best novels may come to envy.

Steve McQueen and Pedro G. Romero form part of this troop. To the extent that, for the Venice Biennale in 2009, they conceived two very different works that unexpectedly complemented each other.

McQueen's consisted of a video. Romero's was a book.

The first depicted the city before the event. The second depicted it in the aftermath. Both established a temporal displacement so they could lead us through Venice without the Biennale.

In *Giardini*, McQueen used the survival of a group of stray dogs to launch his attack on a city whose decadence manages to be cyclically postponed by the Biennale itself.

In *Las correspondencias* (The correspondences), Romero projected an epistolary story into the future, with the goal of building a community out of letters exchanged between neighbors.

The purpose of this community? Not to help people thrive,

---

* An English-language excerpt from "Venice Without the Biennale" appeared as "Verónica Gerber Bicecci and the Language to Come" in the *Centre de Cultura Contemporània de Barcelona (CCCB) Lab* on June 22, 2021.

but simply survive. To offer a respite in the rearguard rather than glory in the vanguard. In this case, to an immigrant in need of collective solidarity enacted in everyday gestures that fall outside the influence of grand doctrines.

Under Blanchot's influence, Romero insisted on stirring up a city that McQueen assumed to be immovable. And while the latter needed to put all his authorship at the service of his critique, Romero's work could only hit home if, on the contrary, it managed to escape its own author.

In the first case, Venice appears as a city in search of an artist. In the second, we find an artist in search of a city.

These two works make visible that, just as there has been an explosion of art in narrative, narrative, too, has launched a bomb into the territory of art. Shockwaves have mercilessly pounded the borders that once separated the two.

All this has precedent. It runs from Bill Viola to Lars Arrhenius by way of Doug Aitken, Marina Abramović, Valérie Mréjen, Stan Douglas, and Chris Cunningham . . .

Some of this has also been sought, in recent decades, by so-called "expanded literature," which has devoted itself to praising writing's spectacular condition and its sociable drift, its connection to new technologies and the performance of them, its links to pop culture and television, to social networks and the exhibitionist side of writers.

Expanded literature is not, strictly speaking, art. (In fact, its natural habitat is a mixed terrain where theatre and cinema, video clip and performance, spoken word and publicity, graffiti and the internet intersect.) But it is undeniably located in a territory previously conquered by art.

If there remained any doubt about this, it would be enough to consider the origin of the very concept that gives it its name. To consider its debt to Rosalind Krauss's "Sculpture in the Expanded Field," an essay that examines how minimalist sculpture, unlike Duchamp's readymade, skipped traditional exhibition spaces and went in search of open spaces, nature itself, to distance itself from the white cube.

(The connection is underpinned by the relationship between Agustín Fernández Mallo—figurehead of this trend in Spain—and the sculptor Robert Smithson, one of the artists Krauss praises.)

And so, expanded literature unites malaise and resignation; criticism of the literary system's status quo and a desire to set fiction down on other foundations; a commitment to crossing genre boundaries and an intention to dissolve them . . .

This is not unprecedented either. A glance at Catalan poetry reveals, for example, that this "expansion" has always been a rite of passage, undergone by Joan Brossa and Enric Casasses by way of Carles Hac Mor and Accidents Polipoètics.

On the other hand, the hype around expanded literature has caused more alarm within conventional narrative than in the territories into which it is actually trespassing. Like an inversion of the butterfly effect, its wingbeat has been stronger at the point of departure than the point of arrival, where we can be sure it has provoked only the merest breeze.

And the fact is that, at this point of arrival—where the Pedro G. Romeros and Steve McQueens and Stan Douglases of the moment await it—a narrative tradition that has its own rules, its own critical framework, its own hierarchy, its own bouquet of vanities, and its own crisis has already been delimited. (From an artistic point of view, a rudimentary book trailer cannot help but look foolish beside a six-screen installation by Doug Aitken.)

Does this mean that the criticism of expanded literature is unjustified? Of course not. Only that, in a world where it's impossible to avoid overdosing on the visual, it's perhaps worth asking ourselves if more images will really help get us out of this impasse.

Perhaps what drives art today is a proliferation not of images but rather of words. Especially those that, faced with the onslaught of the visual, are capable of establishing some sort of imaginary. That's where literary expansion would be most fruitful. In the discovery that it is successful not because its authors resemble artists, but rather because they remain, quite simply, writers. Because they are less *literal* and more *literary*.

Right back in the nineteenth century, writers were already immersed in an intense battle for the recognition of their artistic condition, driven by the desire to reach a dignity like that which was afforded to painters.

"Fiction is an art too!" they would cry out emphatically.

In the twenty-first century, some artists seem to sigh out just the opposite: "Art is a fiction too!"

Or, it's a generator of imaginary worlds, a novelesque endeavor, the endless tale of a symbiosis that's destroying the traditional distance between writers and visual artists, as well as the mutual contempt they used to have for one another.

To prove it, there's the exchange between contemporary writers and the artists emerging from their fiction: Paul Auster and his Maria (or was it Sophie Calle?), Patrick McGrath and his Jack Rathbone, Ignacio Vidal-Folch and his Kasperle, Roberto Bolaño and his Edwin Johns, Álvaro Enrigue and his Sebastián Vaca, Javier Calvo and his Matsuhiro Takei, César Aira and his Karina; Grégoire Bouillier and his Sophie Calle (again), Siri Hustvedt and her Harriet Burden . . .

It's clear that, at some point, these authors were not comfortable with "real" artists, so had no choice but to invent new ones, with no noticeable reduction in the aesthetic usefulness they activate.

"But they're fictional artists," people may complain.

So what? The principal malaise of contemporary culture seems not to come from what we conceive as *fiction* but from what we perceive as *truth*. For all their well-known fantasy, none of these writers has imagined a fiction as colossal as the famous weapons of mass destruction in Iraq, or as Noam Chomsky's claim that the atrocities committed by Pol Pot and the Khmer Rouge in Cambodia were exaggerated by *The New York Times*.

I have recently read four novelists who bring this issue up to date: Vicente Luis Mora (*Fred Cabeza de Vaca*), Alicia Kopf (*Hermano de hielo* [*Brother in Ice*, translated by Mara Faye Lethem]), Verónica Gerber Bicecci (*Conjunto vacío* [*Empty Set*, translated by Christina MacSweeney]), and María Gainza (*El nervio óptico* [*Optic Nerve*, translated by Thomas Bunstead]).

These four books can be read as expeditions into the fairly remote confines of art. Brutal invasions in which art appears not as life itself, as is often said, but rather as an unreachable beyond. In its own way, each book helps us to understand, through reflections on art, what Coetzee defined as the "internal mechanisms" of literature. They are undoubtedly books of fiction. But they also hold the real possibility of changing the dynamic of art in use by accepting it as a literary genre.

If François Truffaut filmed that recurring joke depicting the critic as a failed artist, Vicente Luis Mora turns it around and reconstructs the biography of an artist who triumphs precisely because he abandons his work as a successful critic. The character is Fred Cabeza de Vaca, a twenty-first-century Spanish artist who cements his fame by swinging cynically between the most sublime themes and the most spurious ambitions. It is a cubist portrait of Cabeza de Vaca as an artist with multiple trades—he is a curator, promotor, activist, and opportunist—fully immersed in that amalgamation of art and writing that today defines the Age of the Image.

In *Conjunto vacío*, Verónica Gerber Bicecci reconstructs the fragmentary inheritance of an absence for which she can find no explanation. Hence her "methodology of forgetting" ("ausencia quiere decir olvido," goes the bolero, written not on the shores of the Lethe but on the coast of the Caribbean Sea—"absence means forgetting") in which the tables are turned and art manifests as a distressing experience in which the devil you know is safer than the angel you don't. The empty set of the title is expressed through the narrator's drawings, a school of art that runs in parallel to the school of life. And which returns relentlessly to the question of what is exclusive to the artist in this time when we all perform tasks that previously only an artist was able to undertake: drawing, taking photos, producing images . . .

Alicia Kopf is, like Verónica Gerber Bicecci, an artist as well as a writer, and in *Hermano de hielo* she sends the world a letter that its autistic inhabitants cannot answer in the same frequency. Kopf has something of Vila-Matas in *Exploradores del abismo* (Explorers of the abyss), and, for her, moving through a book involves forgetting aspects of its plot. At the end of the day, reading is nothing more than turning the page. Here, art is a pole as cold as the Arctic. An extreme that fascinates and freezes us at the same time. An ice rink on which, if you want to keep your balance, you have to stick to the choreography.

María Gainza's stories, for their part, evoke Borges's "Funes el memorioso" ("Funes the Memorious," translated by Anthony Kerrigan), only in this case the characters' hypermemory is connected to the impact the work of Alfred de Dreux, Cándido López, Hubert Robert, Axel Amuchástegui, Gustave Courbet, and Aubrey Beardsley have had on them. For Gainza, our fate has already been painted in another era, so life is nothing more than a premonition laid out on an earlier canvas. All we need to do is find the survival kit contained in these works in order to go on surviving.

These four books involve a broad reception of visual images, but they are also equipped against their omnipresent tyranny. They are based on the texture of visual art, but with the intimate objective of turning it into text. They are, we might say, the B-side of many ongoing exhibitions, the core that conceptualists dodge, the first-person resistance in the face of the demands of a social art that has already become the norm.

Faced with the infinite variants of a readymade that no longer gives anything more of itself, these narrative pieces aren't trying to place an object in a museum in order to turn it into an artwork, but to invent that artwork in order to place it in a book, the definitive museum that will determine its survival these days. A "written" painting which, faced with the orgy of good causes, prefers to provide us with what Robert Louis Stevenson would call a "banquet of consequences." An artwork that is not exhibited but that *is* exposed. Given this challenge, literature is no longer content to describe art, but rather devotes itself primarily to constructing it.

One of the most notable antecedents of these tales is *Composition No. 1* by Marc Saporta, published for the first time in 1961. More than a book, it is a "random" object, a set of cards that anyone can shuffle, cut, or deal however they wish.

By doing so, we can put together a story that will always offer an original reading experience, the first of many possible compositions. Its unnumbered pages evoke—as Miguel Ángel Ramos claims in his prologue to the Capitán Swing edition of *Composition No. 1*—Cortázar and Lezama Lima, the *I Ching* and Max Aub's *Juego de cartas* (Card game), Italo Calvino and Julián Ríos. We could also mention Wittgenstein's *Tractatus*, another work that requires the means of a future generation if it is to reach its ideal composition.

Unlike *Rayuela* (*Hopscotch*, translated by Gregory Rabassa), *Composition No. 1* leaves us no map with which to orient ourselves in the territory. Not a single instruction manual with which to build the puzzle it presents. And yet, under the random cadence of its small movements, an invariable line perseveres through the chaos. In that cinematographic shot in which "on both sides of the car, the street files past" and in the adolescent girl who caresses herself, moaning. In the tunnel after a catastrophe and in a tear escaping Helga's eye. In a mouth waiting for a kiss, like something out of a poem by John Donne, and in the policemen playing in their time off. In an accident that allows us to glimpse the story's Big Bang moment and in the protagonist, Dagmar, who manages to paint her own shadow.

Composition No. 1 is also an "órdago" (move) against literature as a trade, and a bet—let's stick with the logic of gaming—on its function as an antidote to routine.

Saporta has been defined as a pioneer of expanded literature, and it's true, because he used mechanisms from other forms of art to make his writing flow. But it's also true he didn't resort to audiovisual jargon to draw attention to his having "crossed the bridge."

Composition No. 1 is built on the risks inherent in gambling, such as debt, betting, and cheating. And on those moments when readers, as gamblers, are allowed to recover so that they may later relapse with a greater roar. So that they may fail better.

This book is also an ambush (but we will learn that too late). A trap built to make sure we will inevitably lose the game. As the author—or his shadow, or an avatar—says, "in the end, the croupiers always win." In this case, the croupier is named Saporta.

Let's turn to someone else who always eventually wins: Michel Houellebecq. These "artistic" themes were already lurking in his *Extension du domaine de la lutte* (*Whatever*, translated by Paul Hammond) and *Les Particules élémentaires* (*Atomised*, translated by Frank Wynne). But it is in *La Carte et le territoire* (*The Map and the Territory*, translated by Gavin Bowd) that he most thoroughly examines the literary site of Contemporary Art.

This leads us to Jed Martin, the protagonist of *La Carte et le territoire*, an artist who has achieved unexpected success with his photographs featuring Michelin road maps. Once he has exhausted this route, the character version of Martin decides it is time for a change. So he returns to painting and portraiture.

The problem is that Martin manages to abandon his formula, but not his success, which multiplies along with a sense of emptiness that makes him detest the world around him.

His Russian lover Olga and his dying father, the gallery owner Franz and Michel Houellebecq himself, who appears as a character in his own novel . . .

Along the way, there is a curious vindication of utopian socialism and a rejection of Le Corbusier; there is criticism of the stars of the commercial art world—Jeff Koons and Damien Hirst—and the dream of a "human" architecture; there is nostalgia for a real life and for the exhausting protocol of success.

And hanging over it all is a question about the irresolvable contradiction between a culture created using a cartographic scale and lives created using a geographic scale.

15. MUSEUMS, MUSES, MUSINGS

Which museums, then, would house these imaginary works? And which ones would laud these artists who were "never real" but "always true"? Well, museums that by day behave like tranquil, cultured places, altars for seclusion where we speak little and touch even less. The same museums that, by night, fill with ghosts and mummies that come to life. Outside, beggars lurk. Inside, atrocious crimes and perfect robberies, ancient truths and orgiastic tendencies . . .

The New York Metropolitan Museum, for example . . .

It houses the last living being on the earth, according to David Markson in *Wittgenstein's Mistress*, as well as Aloysius X. L. Pendergast, the disturbing FBI investigator—half–shadow man, half–David Bowie—conceived by Preston & Child.

These museums, of course, provide fodder for both cult novels and bestsellers, classic and B-series movies, telenovelas and comic books.

The Marx Brothers spent a night at the opera; Mr. Bean's time in the museum was a little more drawn out.

There is news of a secret museum in Barcelona, every inch of which was examined by Ignacio Vidal-Folch (and judged impossible to build). And of Orhan Pamuk's Museum of Innocence (built after the novel was written and more like a museum of naivety).

There are those who promote the idea of turning Guantánamo into a museum (as things stand, unlikely) and we already have a museum island in Abu Dhabi (as things stand, pretty feasible).

Museo de la revolución (Museum of the revolution) is a novel by Martín Kohan that portrays the generation of Argentines who were plunged into armed struggle during the seventies . . .

These examples might easily bring to mind an exhibition titled

The Museum as Muse (MoMA, New York, 1999). In it, the museum was reflected in the work of certain artists; it ceased to be "the house of the muse" and immediately became the muse itself: Thomas Struth and Henri Cartier-Bresson, David Seymour and Marcel Duchamp, Christian Boltanski and Christo . . .

Just as there are imagined artists and imagined works of art, ever since Desiderius Erasmus of Rotterdam, inventing a private museum has been a modern fantasy, a recurring daydream about which we cannot help musing.

If we're being serious, Aby Warburg, André Malraux, and Georges Didi-Huberman come to mind. Being flippant again, amid the ambiguous cobwebs of a siesta, we end up with television series such as *Bones*, hyper-bestsellers such as *The Da Vinci Code*, cartoons such as *The Simpsons*, and unclassifiables such as *Museo Coconut*—so scathing, so authentic, so trashy.

Sometimes, fiction resides in the plot that belongs in "real" museums. At other times, the museum itself turns out to be the fiction.

In fact, when someone talks about a "museo de autor" (single-artist museum or, literally, "author's museum") I tend to attribute it to this type of "authorship" and not to the famous artist-cum-museum director who tends to claim it for themselves. That reaction is due, in part, to an understanding of art as a literary genre. And in part to my disagreement with so-called "artist programing," known in Spanish as "programación de autor" (literally, "author programing").

Because, save for the obvious differences and leaving aside some very notable exceptions, most people in Contemporary Art who consider themselves to be in this category do not program museums because they have previously been "authors" (i.e., artists) but rather have become "authors" because they have directed museums in the past.

There is little to like about the kind of *authorship* that emanates exclusively from *authority*, distant as it is from the kind that comes from the imagination. Or from those unique, personal museums that each of us—Nobel Prize winner or mere mortal—has got stuck in their head like a stubborn daydream.

The kind of authority I'm rejecting here usually finds its place in the insufferable jargon that has stagnated criticism for a while now. This vocabulary elevates a type of critic who cannot always be considered a "writer"—there are those who go on stubbornly pounding at language—but who can neither be treated as "unpublished." In fact, their output might be prolific: disseminated through a network of magazines, event summaries, websites, and other publications, almost always contingent on the work of artists or circling in their orbit.

The star of this authority without authorship is, without doubt, the catalogue. There is no more generous support, no better host for the incomprehensible gibberish of this sect that aims to be as impenetrable as a dead language.

The problem is that catalogues—with a couple of exceptions—are not read outside of the art cult. And the few that are read unfortunately live short lives beyond the specific timeframe of an exhibition (which is why publishers insist, desperately, on publishing them alongside the exhibition, trying to grab at their handful of sales during the period when the show is still active).

The vicissitudes of publishing are not the only obstacles catalogues have to overcome. Their dimensions—size XL, usually—are not a minor problem, nor is their weight, nor how uncomfortable they are to read.

Out of proportion. Overweight. Hard to handle . . .

The catalogue is to the book world what cetaceans are to the marine world. An exaggerated beast that, at some point or other, most mortals (as well as some immortals) will find themselves consumed by—assuming it does not first go extinct.

Assuming we exclude furious anti-collectors (men who have no regrets when it comes to throwing out or burning books) such as Onetti, or Vázquez Montalbán's detective Carvalho, the truth is that this dilemma has made martyrs out of many.

One well-known case is that of Gabriel Zaid, who gave an account of this anxiety in *Los demasiados libros* (*So Many Books*, translated by Natasha Wimmer). As far as Zaid was concerned, life had given him more books than he was due, just as life had given Desi Arnaz more girls than he was due—in fact, *Too Many Girls* was the title of the 1939 musical that made him famous, before the heyday of *I Love Lucy*, the famous television series in which he starred for a decade alongside Lucille Ball.

"I'm the *I* in *I Love Lucy*."

But back to Gabriel Zaid and that volume of his that examines Luther, Herodotus, and Italo Calvino's bookish agony. To that long essay of his tracing both the horrors of the publishing industry and the no less horrifying litanies of friends who, not content just to give us books, also demand—sometimes to our faces—to hear our opinion on them.

The fact is that, somewhere between an archivist and a collector, the cataloguer is, in large part, a "discontinuer." And the fact is that this function is not exclusively limited to judging the sell-by date of some edition or other, or to proving the impossibility of preserving such and such a copy. It also involves taking a stance on that medium itself—the catalogue, both now and in the future, and also—and this should be said as discreetly as possible—on the art memoir itself.

Are catalogues, like everything else, in crisis? The very protagonists of this medium seem to think so. To the extent that, in the last few years, there has been a clear fervor among curators and artists for making catalogues "that look like books."

"Make it look like an accident."

This goes both for the format and for a hidden will to transcend, to reach beyond the ephemeral life of exhibitions.

To me, however, the hecatomb isn't so obvious. It's true that a catalogue that functions only as a sumptuous object will be condemned to being just that, an object; which is to say, condemned to itself. But it's also true—if we have learned anything from Aby Warburg or Georges Didi-Huberman—that catalogues have infinite possibilities available to them.

These days, when we can enjoy virtual exhibitions and museums, it's easy to see how much catalogues could give of themselves without having to take up meters of physical space. As part of this expansive wave, you could feasibly have detailed catalogues of artistic processes, catalogues that would allow us to see projects in real time, interactive archives of images, comments, and footnotes, whole chapters that include the critique unleashed by the exhibition.

In this age of crowdfunding and other manifestations of the ancient—and very artistic—custom of a whip-round, not even printing would be a problem. (It could be done à la carte, in various versions, and in different scales of print on demand.)

And as for old or unmanageably large catalogues, there is another simple yet practical alternative: Turn them into PDFs before turning them in to the firing squad.

PART V

One or the Other—Immortal or Contemporary (How to Write an Epitaph for Contemporary Art)

16. FUKUYAMA, CONTEMPORARY ARTIST

November 9, 2012, marked the twenty-third anniversary of the fall of the Berlin Wall. It was a Friday, and Francis Fukuyama "celebrated" the anniversary by dropping into his local Sears. He went directly to the tool department.

While he was there, he shared a tweet with his then 13,000 followers: "I've been feeling nostalgic." Thus spoke @FukuyamaFrancis.

November 9.
Sears.
Tools.
Nostalgia . . .

Just the combination to incite, in the end-of-history guy—the man who foretold us a long and happy boredom, the last man's man—a fit of Luddism. An unassuageable longing for the age of manual labor and, who knows, perhaps even for the Cold War itself. A flash of melancholy for Sunday DIY before a barbecue, helping his dad mend something on the porch.

Fukuyama himself, lost in the memory of a time when capitalism still had a bucolic tint. With its Woolworth's milkshakes, a newspaper flying over the gate, and a liter of milk on the doorstep.

The man for whom—to paraphrase Lennon on Elvis—"nothing existed" before 1989, lets himself be struck down, one November 9, by the memory of those times when everything, even history and the future, existed purely for the sake of existing. That time when twenty-first-century humans still organized themselves around the future, as primitive humans had sat around the fire. A remote period in which goals were still valid and achieving them was even less important than pursuing them. That long ago era when parents, in short, worked, went to war, or brought on a revolution so their children could have better futures than they did.

All this suddenly occurred to Fukuyama on the anniversary of the event that had led him to proclaim the end of history (in which, if you really think about it, he wasn't betting on the end of the past but rather on the end of the future).

That morning, Fukuyama also gave himself, as it were, "a reverse résumé." Just like those catalogues that propose a journey to the artist's primordial purity. That day, this friendly, lighthearted neocon behaved like a contemporary artist.

Throwing himself into the moonwalk—like a Michael Jackson impersonator prepared to delight customers in the department store, so archetypal of old capitalism—he slipped into the past while convincing his audience that he was headed for tomorrow.

It cannot be a coincidence that both history—charged with rewriting the facts—and art—busy aestheticizing them—conceive the contemporary as an eternal dimension. (Lenin himself had identified the contemporary era, born with the Communist Revolution, as the last and definitive season in the human calendar.)

Nor can it be a coincidence that this history after "the end of history," foisted upon us by Fukuyama in the place of Leninist contemporaneity, ran so placidly alongside the art after "the end of art" into which Arthur Danto managed to accommodate contemporary aesthetics.

These two eternities—one applauded by the Right and the other by the Left—have in common the sublimation of a present continuous; proven by that prolonged vigil in which we continue to subsist, accustomed as we are to the art of shrouding bodies that were easier to kill than they are to bury.

And so, the funereal deluge has from time to time come to notify us of the deaths of communism and art, history and man, truth and ideologies. For this reason, our contemporaneity could only aspire to the eternity of a prefixed existence—post-everything—and to the kind of posterity without a future that Peter Sloterdijk christened "the age of the epilogue."

In this vein, Luis Goytisolo can decree the end of the novel just as Fukuyama had proven the end of history. And Milan Kundera can consider Francis Bacon to be "the last painter" just as Roger Caillois can define Picasso as the "great liquidator" of art. And that's not forgetting Theodor W. Adorno, who denied the possibility of poetry after Auschwitz . . .

However, all these obituaries, appearing on the edge of life, run headlong into our dilemma of inhabitants in survival. Into this stubborn rearguard from which we extend life by other less certain means. Like zombies forced to make art after Picasso, paintings after Bacon, history beyond communism.

The challenge of this rearguard culture will no longer consist, then, of informing people—one death at a time—of all the things that are ending, but on whose carrion we continue to feed, insisting on a forensic worship that no longer frightens nor saddens but simply bores us.

Perhaps the time has come to name our experience positively again, and that will involve not only burying the corpses that serve as its ballast, but also ceasing to exploit their ghosts.

In this rearguard zone, perhaps we do not deserve a *better* future, but we do deserve something of one. Not a future tomorrow, but this future that is happening today and that is used up, over and over again, in the name of an unperishable contemporaneity incapable of assimilating its end.

It's not an easy task.

Along with the fall of communism, this preservation of contemporaneity has been linked to the upheaval caused by the rise in new technologies (that movement, as we have seen, from one PC, the Partido Comunista, to another, the Personal Computer).

To the point that many now quietly accept the rise of a superstition. The one that claims that, in order to reach the future, we will no longer need to take humanistic paths, because we have machines now, which are much less fallible.

These two major events—the collapse of communism and the apotheosis of technology—have also nourished the foundations of a contemporaneity that has often been depicted as a visit to the apocalypse.

A quick survey of science fiction from recent years is enough to confirm that the more realistic the plot—technology means that any hyperreal effect can be achieved on screen—the less habitable the future, populated by cyborgs, deep freezes, posthuman sects, and environmental catastrophes. In almost all of them, humanity is a thing of the past. And not only because it cannot find a place for itself in what is to come, but also because it can neither improve itself nor that future in which humans are destined to eke out their survival like vagabonds.

The victory of contemporaneity over the future—with that regressive loop that marks the passage of time in the art résumé—is complicated even further when we find it is not only connected to the end of the Soviet Empire but also to a growing certainty about the mortality of capitalism, which Fukuyama had proclaimed—just as Lenin did with communism—to be an eternal entity.

Eudald Carbonell is convinced, on this point, that capitalism too will disappear during the twenty-first century. The curious thing is that it will not happen through revolution but rather due to a process of "thermal death."

This is how he explains it in his memoir, with its contradictory title: *El arqueólogo y el futuro* (The archaeologist and the future).

That an archaeologist would offer us tips for how to inhabit tomorrow (which is our *today*) could be considered a bad joke by a scientist blinded by his own interests. Among other things, because he is neither speaking in metaphorical terms, nor in the utopian vein that Fredric Jameson adopts in his archaeologies of the future, nor the somewhat more tragic tone of Foucault's archaeology of knowledge.

Nothing of the sort.

This archaeologist speaks from the experience of someone who has dedicated himself to boring into the earth, convinced that the mystery of our future will not be found in a game of anticipation but rather in an exercise in excavation. Because it's not that the future has been postponed, but rather hidden: We need to pull it up into the light rather than waiting for it.

If we look at things this way, what we need is a revised history of the future. Or at least a "recollection of things to come," as it says on the doorway of a Mexican tavern. And if the future is already here, if it's *this* that we are living, it's even more important to dig into the different layers covering its surface rather than continuing to coat it with layers of varnish.

It's not an easy task; nobody said it would be.

And not only because of how hard it is to talk about what's to come when the refrain "we have no future" echoes around us. It's hard to call for the future in a desert populated by people who have been denied one.

17. THE ART TO COME

What can art do in this situation, other than wrap itself up in a contemporaneity that precludes an expiration date?

It's time to return to Michaux . . .

The mortality of Contemporary Art (which has cashed in on the death of art and of the author) was already implicit in that phrase of his that has crept here all the way from the beginning of this book.

The one that defines the artist as someone who always leaves a trace.

The paradox lies in the fact that today the feverish desire to leave one's mark has affected culture more profoundly than ever before. To prove it, we've got Facebook and Twitter, Instagram and LinkedIn; the tens of millions of blogs and their inevitable demand that we all bear witness to the most minimal experience.

Our problem, monsieur Michaux, is that no one can resist leaving a trace anymore. What's changed is that the pressure to do so has stopped being exclusive to artists. So, either your definition has expired, or the old tasks involved in art no longer require artists to execute them.

If we look at history, we have to admit that *this*, specifically, is what utopia was for young Marx. Creative leisure in which common people, during their spare time, could hunt, fish, rear cattle, or criticize without needing to be exclusively "a hunter, fisherman, shepherd, or critic."

Now that we're in paradise, it's worth remembering the even more radical step Joseph Beuys took when he announced that, by performing these tasks, everyone could consider themselves artists, rather than mere mortals who acted "as if they were."

Within this realm of possibility, art's malaise would no longer lie in its invisibility or mystery, but rather in its ubiquity, and in that multitude of traces cramming into the horror of emptiness that governs our every step, but over which art has already lost the monopoly of exclusivity.

All this speaks of a generalized experience in which people mount a permanent exhibition of their daily epic. After all, why wait years for a museum to take your work, or a publishing house, or some other temple of culture, if you have at hand the immediate possibility of being the curator of your own exhibition, the publisher of your own novel, or the DJ of your own music?

With the support of any platform—With a Little Help from My *Facebook*—we can mount a daily display of the confessional community emerging in the Age of the Image. This culture of millennials nourished on a do-it-yourself attitude, for whom Facebook has ended up being a mix of Beuys's dream (we are all artists) and Duchamp's conviction (everything can be made into art).

Faced with this apotheosis of traces and signs, it's possible to intuit, against the tide, a change in the condition of the artist Michaux recognized. Meaning that perhaps now the most interesting artists are those who hasten to erase their traces. Those who won't question us obsessively and who, while they're at it, will cover their tracks so we cannot find them.

Like Wolf in *Pulp Fiction*, but with some Photoshop training . . .

In the face of the planned obsolescence of new technologies, there is the planned immortality of Contemporary Art. Especially in a world that runs on addiction. And whose subjugation means "to discover consumers, to excite their appetites and create in them fictitious needs."

This phrase belongs to a suicidal revolutionary. But didn't appear on any 15-M or Occupy Wall Street banners, nor was it cried out by any of their speakers in order to whip the campers into a rage. It's actually a century and a half old, written by Paul Lafargue in his *Le Droit à la paresse* (*The Right to Be Lazy*, translated by Charles Kerr).

Like the rest of that book, this message was directed at the future. At days such as these when, the more we consume, the quicker it is proven that everything around us has an expiry date: cars and medicines, buildings and computers, beliefs and illusions, secrets and lies, husbands and wives.

Everything has to be replaced. And the sooner the better. It matters little that, in most cases, the objects or people being substituted—including a husband or two—still retain their faculties and perform their "services" reasonably well.

The fact is that we do not produce artefacts or ideas, machinery or works of art that compete on the market of durability, but rather on the market of transience. In this empire of the ephemeral—formerly the exclusive domain of fashion—terms such as "expiration date" and "obsolescence" are not entirely synonymous.

The newer and more sophisticated the gadget, the quicker it is likely to be declared obsolete.

A situation that doesn't always coincide with a decline in its operability.

It isn't their decrepitude that takes our "toys" out of circulation, but rather a drive to replace them imposed by the addictive dynamics of their consumption.

Now settled into the future (in this twenty-first century that has seen almost all science fiction's fantasies come to pass), our nostalgia suffers a displacement, so to speak: It is not directed towards the past but rather towards a present that seems to burn out at the same speed as the objects it comprises.

18. ONE OR THE OTHER: IMMORTAL OR CONTEMPORARY

In what we have seen of the twenty-first century so far, terror—whether from Islamists, narcos, or Western illuminati—may continue to set the future on fire, but it hardly ever burns books anymore.

It still burns people, sure; or rather, it decapitates them.

But its cultural psychosis prefers to target statues, the open-air museums that still abound in the East, architecture itself.

In short, any artistic heritage or human event that reminds us of the relevance of civilization.

Few words, but many images. That seems to be the motto.

Given that this horror tends to alternate demolition with filmic construction—video-terrorism—the ardent, implausible speeches of other times have been discarded or reduced. Those pompous communiqués, the messianic manifestos in the vein of the Unabomber or bin Laden, seem to wane in the face of the visual impact brutality can offer us directly.

The new imaginary of violence seems to reaffirm the majority criterion according to which, thanks to the expansion of visual culture, art has destroyed literature—has crushed it under the weight of images.

It's possible, however, to test the truth of this widespread superstition. And to intuit, contrary to what everyone's been lamenting, that the apotheosis of images will end up being irrefutable proof of literature's survival.

It's true that bookshops are closing, but also that books themselves remain. We cannot deny that the way we access text is changing, but neither can we refute evidence that people still read.

This was precisely what Aldous Huxley warned, more than six decades ago, in an essay: "If My Library Burned Tonight."

There and then, Huxley argued that the burning of a library would affect collectors and even fetishists—by destroying incunabula, first editions, annotated volumes—but that in the strict terms of literary death, burning down a library was a pointless act given that the books could always be owned and read again.

To see this from a different angle, let's suppose all the museums burned down today. Despite all the technology at their disposition for the reproduction of artwork, we cannot deny that a large part of the art would disappear among the ashes.

Not least because one of the museum genres par excellence—the exhibition—would be unceremoniously incinerated.

Someone will object that you will always be able to see a van Gogh or a Picasso reproduced to the millimeter on any computer screen you like. But we can agree that there is no digital technique, however precise it may be, that will allow us to appreciate the *Perro semihundido* (*The Dog*) exactly as Goya painted it, whereas an e-book *can* allow us to enjoy *Hamlet* exactly as Shakespeare wrote it.

The profusion of visual images has not strengthened art. On the contrary, it has made it spontaneously combust. The curious thing, though, is that we insist on not seeing writing's discreet victory, just as we do not seem to notice the evident defeat of art.

Or perhaps we just prefer, as ever, not to look at ourselves in the mirror.

This phobia of looking at oneself in the mirror is revealed in the following detail: When the art world—the Sector—comes together, it does so to discuss union matters, managers, or political battles, but almost never to talk about art, its own procedures, or its place in the world. On the other hand, writers—harassed, perhaps, by that hecatomb that has been made so much of—tend to talk about literature. Perhaps because they are as distressed about the predicted end of their world as artists seem euphoric about the end of theirs.

This is probably all owing to a distinction that, in the light of these times, is no small thing: While contemporary literature has long been involved in resistance, Contemporary Art still has revolution as its guiding star. And it's true that, while we resist for ourselves, we tend to make revolution for others. Just as we can live for others but can only survive for ourselves.

This, and nothing else, is the goal of expanded literature, which is not the result of an exultant colonization of other fields, but rather an attempt to survive after the implosion of its own: through images, new technologies, and the transformation of reading and the book itself as the ideal, centuries-old conduit for knowledge. Rather than a relaxed choice, this expansion responds to an agonizing need. And more than a cheerful act of cross-dressing, it is a sign of anxious, often mandatory adaptation in the face of its arrival at the precipice.

As far as artists are concerned, it's likely that their obsession with ideology, documentation, social activism, fashion, advertising, and political demands has led them to neglect the most important task this Age of the Image had in store for them. And, despite being faced with the definitive triumph of their media—photography, video, the internet's innumerable visual possibilities—they have not succeeded in becoming the intellectuals of an era in which, increasingly, knowledge is distributed through visual aids.

Art's malaise is not borne of its difficulty, but rather of its feasibility. And of the overwhelming presence of media that used to belong to artists alone and which today keep a record, second by second, of the infinite traces that testify to the immense horror of emptiness that governs contemporary culture. The problem is that, once Beuys's dream has been fulfilled, what was proposed as therapy begins to manifest as a virus; it's the disease, rather than the cure, that is spreading.

Hence that artistic lifestyle without art, aesthetics without poetics, that unstoppable drive to exhibit without the need for a museum.

Contemporary creators have placed such high bets on the expansion of art that they have been unable to control the bits and pieces left hanging in a galaxy, barely functioning as an image bank, despite art specifically being called on to generate imaginaries. That spread, in part, has cleared the museum but not set it alight. Rather, it has frozen it, keeping it a neutral space still known as the white cube (or was it an ice cube?).

On this point, we could put the same question to artists that Klossowski asked when thinking about Nietzsche: Is the philosopher still possible today? Are they necessary?

Let's return to Rosalind Krauss and "Sculpture in the Expanded Field." And to art's leap beyond its own confines, when it sought an anthropological dimension from which to safeguard the human scale.

A human had already walked on the moon and Kubrick had released *2001: A Space Odyssey*. Paul Virilio had already talked about the aesthetics of disappearance and Nam June Paik had founded video art. That same year, 1979, Lyotard published *La Condition postmoderne* (*The Postmodern Condition*, translated by Geoff Bennington and Brian Massumi). But Ana Mendieta and Robert Smithson preferred to go back in time to unsettle the hierarchies of the Western world, examine the persistence of humanism prior to modern life, and investigate that moment before culture began to circulate as a good.

It's obvious that these artists made use of the latest technology available to them, but it wasn't that which determined their curiosity. It wasn't the technological revolution, nor even politics, that spurred them on—although they were no strangers to either—but rather a human resistance, perhaps too human for us to digest it completely these days.

Art was experiencing a discomfort that it could no longer resolve within its limits and so, as Engels pointed out about capital, it had no choice but to expand or die.

Today, however, its dilemma involves the opposite sequence: It must either shrink or it will disappear.

Huxley gave an account of this shrinking; Huxley for whom there was no way out other than to reduce the abundance that was leading the art of his time towards mediocrity, just as mediocrity was leading literature, as it were, towards abundance. So the fire not only seemed to him a lesser evil; in truth, he secretly understood it as a necessary evil. And not so much because it would destroy the library or the museum, but because, thanks to that fire, we would be forced to rebuild their order and scale of values.

That's why there's barely any point talking about Contemporary Art these days.

The term "contemporary" has come to mean nothing at all. Mainly because contemporaneity doesn't denote a temporal, tangible magnitude but rather one that is "professional" and indefinite. It's a prop for our presumption of eternity, where Lenin, Fukuyama, and Arthur Danto meet, and from which the death of art appears as one of the most profitable aesthetic genres: one that imagines the end of art as a fine art in itself.

For this reason, what we urgently need is not avant-garde art but rather an art of the rearguard. An art which, faced with the impossibility of amalgamating with life, at least achieves a fruitful fusion with survival.

Seen thus, it's possible to dream of hidden artworks that manage to validate how Huxley described this precise moment: If we had time to think of anything but the economic crisis, we'd realize we are also in the grip of an aesthetic and intellectual crisis.

This is what rearguard work would be about, functioning as a hazmat suit to protect against the igneous ability of an era that burns for multiplication, abundance, overexploitation, quantity, countless numbers, and the definitive triumph of the possible over the necessary.

An act of usurpation that would replace *exhibition* with *inhibition*.

Just when the contemporary has exhausted its substantive condition to become an adjective for the artistic experience, we can calculate the non-existent continuity of what we continue to call Contemporary Art.

The capital letters, too, imply another intensity that speaks of a System, a World, a circumstance in which the contemporary defines not a time but rather a sphere, not a genre but rather a formula. Not a condition but rather an institution.

From this new perspective, it's worth calling into question art's compulsion to colonize everything that's "more than art." That expansive phase in which the ivory tower is blown up so that artists can experience real problems. The expansion that has led art to fire out a centrifugal energy. Including its abundant discussion of anything at all in exchange for silence about itself.

This involves repeating Duchamp's awakening inside an infinite readymade and Fukuyama's finite department stores. Ready for us to choose caviar in one and decry poverty in the other. Knowing that in one we are allowed to apply Leninism as a fine art while in the other we apply ourselves to the fine art of neoliberalism. In one, believing ourselves to be the masses. In the other, to be amassers of wealth.

The more art is charged with good intentions, the more it can behave like a critical phase of the mechanisms it condemns. And the more abundant its outward journey appears—towards society, politics, advertising, literature, the media, technology, or activism—the poorer its return to its usual lair—the museum and its variants—where it returns again and again, as the incredible diminishing art.

The greater its leading role in the Age of the Image, the more art needs an imaginary of its own that allows it to stop using thought as an imported genre. Especially at this time when events, having first occurred as tragedy and then as farce, now occur as aesthetics.

These contradictions may help us stop decreeing, à la Sartre, that "hell is other people." As though we weren't part of the world we are judging.

We have to get away from art for the sake of the art industry; for the sake of the fans, the likes, and the rounds of applause that this contemporary sectarianism lavishes on itself, as committed to criticism as it is scornful of self-criticism.

To abandon this zombie drift in which, in order to save the "contemporary" we have decided to connect it with the infinite when perhaps the most important thing is not to put art at the service of its time—as (bad) political art does—but rather to remove it from servitude to this era.

Who knows, perhaps this way, after decades of profiting off its own farewell, Contemporary Art might find its final virtue by learning, at last, to take its leave.

ILLUSTRATION CREDITS

All the photographs, taken by Lali Piera, are from *Teoría de la retaguardia manipulada* (Theory of the manipulated rearguard), a series by Guillem Nadal dedicated to this book.